William Robert Guilfoyle, Victoria Royal Botanic Gardens

Annual Report On The Melbourne Botanic Gardens

William Robert Guilfoyle, Victoria Royal Botanic Gardens

Annual Report On The Melbourne Botanic Gardens

ISBN/EAN: 9783741125379

Manufactured in Europe, USA, Canada, Australia, Japa

Cover: Foto ©Klaus-Uwe Gerhardt /pixelio.de

Manufactured and distributed by brebook publishing software
(www.brebook.com)

William Robert Guilfoyle, Victoria Royal Botanic Gardens

Annual Report On The Melbourne Botanic Gardens

Melbourne Botanic and Domain Gardens, Victoria.

ANNUAL REPORT

ON THE

MELBOURNE BOTANIC GARDENS,

BY

W. R. GUILFOYLE, F.L.S., C.M.R.B.S., London,

DIRECTOR.

WITH PLAN OF THE GARDEN.

MELBOURNE:
BY AUTHORITY, GEORGE SKINNER, ACTING GOVERNMENT PRINTER.

M DCCC LXXV.

ANNUAL REPORT

BOTANIC AND DOMAIN GARDENS, MELBOURNE.

TO THE HONORABLE JAMES JOSEPH CASEY, M.P., MINISTER OF
LANDS AND AGRICULTURE, ETC., ETC.

SIR,

In the Botanic Gardens during the year some important steps were
taken in advancing my plan for the general laying out of the grounds;
and the production of landscape effect; such as the effacing of narrow
and useless walks, the substitution of broad sweeping ones in their stead,
formation of lawns, transplanting of trees, clearing the Lagoon, and other
necessary works.

During the year there were introduced into the Botanic Garden 1,122
new species and 1,272 new varieties of plants.

The islands in the Lagoon near Princes Bridge were planted with
trees and sown with grass seed early in the year.

These islands have not been formed after the style which I think
objects of the kind should have in such a sheet of water—they are too
flat, and far too numerous. However I did the best I could with them
so far as planting was concerned. They were attended to periodically
throughout the year. I have again to acknowledge the courtesy of the
Colonel Commandant in giving me the use of the pontoon raft to convey
very large trees to the islands. Ceanothus, Pittosporum, Buddleya,
Laurustinus, Cupressus, Dracæna, Arundo donax, Gynerium, Pinus,
Poplars, Agave, Araucaria, &c., &c., were also suitably planted on them.
The Fern Gully in the Botanic Garden made great progress during the
year. On one of the Lake islands a number of redundant cypresses,
which marred the view, were removed, Dracænas, Gynerium, &c., being
substituted. Aralia papyrifera, Aloe arborescens, Dracæna, &c., were
removed from other portions of the Garden, and placed in groups in
various parts of the ground. A number of choice ferns from New
Zealand and the State Nursery at Mount Macedon were also planted.
A new walk branching towards the Fern Gully, was laid out, the
lawns round the Director's house completed and planted with Azaleas,

A 2

Camellias, and other choice shrubs. These were placed in this enclosure for protection, to be propagated extensively for the decoration of other parts of the ground. The new borders received large numbers of flowering plants—Chrysanthemums, Pelargoniums, Oxalis, Mesembry-anthemums, Lantanas, and bulbs—the latter such as Amaryllis, Narcissus, Nerine, Ixia, Iris, &c. In the Fern Gully two large specimens of Ficus, 20 feet high, were placed; a Loquat 8 feet high ; a Balfouria, 25 feet ; Gleditschias and Ailanthus, 25 feet, and other large trees, to shade the ferns. Early in August the new collection of camellias in pots was planted out. The botanical collection including many hundred species of dried plants, carpological specimens, &c., was formed. It is being continually added to, and will ultimately prove exceedingly useful for reference. A new walk at the back of the Fern Gully was made ; and spaces cut through the Melaleuca scrub, to afford glimpses of the Lake scenery. A temporary strip of ground for a collection of grasses was prepared and planted, the grasses being duly labelled.

In December the rose stocks in the Garden were budded, and an Amaryllidaceous bed was formed and planted. From Bishopscourt, 1,093 plants were received, having been purchased by the Public Works Department some time previously. Throughout the warm weather, every attention was paid to maintain an effective floral display in the Gardens; in fact, this was done throughout the year as far as possible, a succession of bloom suitable to the seasons being kept up, the roses at this time receiving special attention, while the annuals and other flowering shrubs were very plentiful. Lawns and groups were in many places substituted for long formal beds, containing dry, poor soil ; and large quantities of earth, were carted to improve the condition of the beds. The soil of the Palm House Lawn is very poor and I intend this season to topdress it with rich mould to make the grass exuberant in growth.

In the Domain, large trees were planted, and drained by cutting long open drains in the stiff clay, the rockery planted with Agave, Aloe, Gasteria, Sedum, Mesembryanthemum, and other suitable plants.

The Rhododendrons from the American Garden were removed to the nurseries in the Botanic Garden. Many large trees—Pinus, Araucaria and Cupressus, were transplanted from the Botanic Garden ; new walks were formed ; the planting of the Fern Gully in the private grounds continued, and the pruning in the orchard accomplished. About 60 elms were planted along the edge of the Domain, next the St. Kilda Road, Palms, &c., were also planted, with other miscellaneous plants suitable for effective grouping. Much danger was caused during the hot weather by thoughtless persons smoking, the long grass being like tinder, and great probability existing of fires being caused whereby valuable trees, the growth of years, might be scorched and destroyed in a few minutes. This is a matter which necessitates extreme vigilance. A walk was marked out leading from the house formerly occupied by General Chute; the Domain nursery was extended and trenched, and will be further extended, as I require the space, for hardy trees, and shrubs to ornament Government House Gardens.

The back road and approaches to the Government House, carriage stand, stables, &c., were lined out, and buffalo grass planted along the walk encircling the lawn in front of it. A large quantity of plants have from time to time been lent from the Gardens for exhibition at the various horticultural shows; but I regret to say, the knocking about they receive in most cases takes a long time to repair, thus depriving the public of the display they have a right to expect. It would be a good plan if a space of ground were available in the Garden for shows of this kind to be held, as it would be the only remedy for the evil complained of. Having given above an outline of the work performed in the Botanic and Domain Gardens during the year, I subjoin some general remarks on the progress made.

The incidents occurring, and the ideas suggested, during the working management of these grounds for the past year were of course very numerous. A thorough system of drainage having been carried out in the Botanic Garden, its beneficial effect on plants previously suffering from excessive moisture was markedly apparent. Trees, when planted in stiff clay soils—as I fear is the case with the subsoil of our Botanic Garden, and many parts of the Domain also—are liable to have their growth retarded, if they are not altogether killed; from the fact that the soil not being porous, the moisture given cannot escape, and therefore becomes stagnant. Hence my first step was to thoroughly drain the Garden, the superfluous water being conveyed by drainage into the Lagoon. This sheet of water has been cleared and kept free from weeds, and now presents an attractive appearance. The water-fowl on it are doing well; five cygnets, recently hatched on one of the islands, are all thriving. The gold fish are largely preyed upon by the cormorants; the Mayor of Melbourne having kindly given permission for the use of firearms, numbers of these rapacious birds were shot, but this seemed only to encourage others to fill the places of the defunct. The scenery around the lagoon has been materially improved by planting trees, especially those of pendulous, weeping habit, on its banks.

The point of commencement selected by me to carry out the remodelling of the Garden was the part immediately surrounding the Director's house, and therefore well watched. This ground was then enclosed, and choice camellias, azaleas, &c., thus protected from the attacks of larrikins. Previously, there were only two varieties of camellia in the Garden; we have now eighty of the choicest sorts. My intention is to keep down the buds until the plants attain a good size. A large number of new azaleas and roses have also been planted. This portion of the Garden is now complete, showing broad lawns, thickly clothed with verdant grass, interspersed with varied picturesque groups, single specimens, and clumps of trees, and beds of brilliant flowers. This, as a finished portion of the design (of course allowing time for the plants to come to maturity) affords some idea of the intended general effect; and has attracted great attention from the public. The proportion of visitors during the past year has been greatly in excess of the previous one; and from the notice generally taken of the alterations and the show of plants and flowers, it is evident that public interest in the progress of the Garden has much increased. I trust that this will continue; and that the grounds will not only prove useful to the botanical student, but also become a favourite " lung of Melbourne " for those interested in beautiful scenery simply for its attractions.

Much was accomplished near the Director's house during the year in forming an entrance to the grounds by a broad, sweeping walk, which in its curvatures gives shape to a triangular bed of dwarf shrubs, in no way obstructing the view afforded of the spacious and picturesque sheet of water which is the grand feature of the Garden. Immediately below this bed a gently sloping lawn, descending to the edge of the Lagoon is being substituted for the narrow walks which in this part formerly crossed each other, and upon the sides of which grew so many common indigenous trees plentifully represented in other portions of the grounds. In this spot formerly stood the emu pens and monkey cages.

Before commencing to obliterate these walks, I began to form groups in which tropical and sub-tropical plants will eventually be the prominent features. Through these, dispersed with a view to landscape effect, glimpses will be afforded of the clear lake, studded with islands, the careful plantation of which will materially add to the diversity and charm of the landscape. On this lawn I am endeavouring to imitate as much as possible natural tropical scenery. Most people will agree with me that it is far easier to lay out a *new* garden than to remodel one in which blunders have been made, not only in the general laying out but in the planting of trees and shrubs. It should always I think be the aim of those in charge of public gardens—not to reproduce vegetation which may be seen in other portions of such gardens—but to bring before the public, in special spots scenes of beauty not to be found elsewhere, by representing plants of a different character to those more or less common to the locality. Thus, on the lawn previously mentioned I have partly created groups of scenery which being uncommon will add to the attractions of the Garden. The iron fence at present round this lawn is merely a temporary one to prevent people walking over the ground until the buffalo grass with which it is being planted has time to spread and form a good sward.

In my first report I mentioned this grass as splendidly adapted for edgings and lawns, and a visit to the Garden will prove that no other grass can excel—even if it can equal—the buffalo grass in clothing the surface on which it is planted for ornamental purposes.

I intend making a good display (when means are available) upon that portion of the Garden known as the Melaleuca Scrub, on the north-eastern side, near the Botanical Bridge. This is an eligible spot for such plants as Magnolias, Azaleas, Rhododendrons, Camellias, Ericas, Arbutus, Palms, Cycads, tree ferns, Hydrangeas and a host of others. It can be rendered one of the most attractive portions of the grounds. From here by a little art the lake surrounding it might be made to represent a beautifully winding river; and glimpses might be occasionally had of the higher parts of the Garden.

On the lawn near the Palm house, I am planting specimens of vegetation of the more temperate zone, such as numerous species of Abies (single specimens), many Pines, species of Oak, Poplar, Plane, Ash, Larch, Elm, Chestnut, Walnut, Beech, Holly, &c., &c., which cannot fail to prove interesting to the large number of visitors in whose recollections of "Home" these trees will find a familiar place. To pass gradually from the sub-tropical into the more temperate zone is in my opinion one of the chief objects to be kept in view in creating a public Botanic Garden. I am also adopting a rule of placing various species of any particular genus—such for instance as the Berberis, Veronica, Salvia, Bouvardia, &c., &c., in order one after the other, so that the visitor who may take an interest in knowing or learning the names and nature of the kinds of any particular tribe may see them at a glance. I intend creating a Palmetum at the head of the fern gully, as the spot chosen will not only be suitable, being sheltered, but appropriate in carrying out the idea of changing the character of the vegetation in various parts of the Garden. At the same time, I have already paid considerable attention during the past year to the cultivation of this most interesting and charming tribe of plants, with which I have long been associated in the course of my travels in the tropics. There are few of these graceful plants to be met with in the public gardens of this city ; and yet a great many kinds would thrive here quite as well as in the other colonies. In order that I might the more readily naturalise here our Australian beautiful kinds—the Livistonia or Corypha Australis, and Seaforthia elegans (two species which are found far north in Queensland as well as in the cooler clime of Illawarra, 50 or 60 miles south of Sydney), I introduced a quantity of seeds from the latter locality, and they have germinated freely. We have now some hundreds of nice plants, which shall be liberally used in the decoration of the Government House grounds as well as the Botanic Garden. The lifting of large trees at certain seasons has generally been looked upon as a great mistake. Many opinions have been expressed to me that to lift a large tree say 35 or 40 feet in height—is utterly impossible in this colony. When doing so in these gardens during the past year, I have often been told that they would perish. I need only instance the fact that large Norfolk Island pines, Bunya Bunya, Pinus insignis, Cupressus macrocarpa, Brachychiton populneum, B. Acerifolium, Cordylines, and many others, averaging from 20 to 30 feet in height, have been lifted with perfect success during a season hitherto generally considered unfavorable for such operations. In my opinion, and it has been the secret of my success—in such a changeable climate as this, when a plant is suddenly forced into active growth, and this growth is observed in its commencement the evergreen and even the deciduous plant, may be lifted if due precautions are taken in digging far enough away from the tree, so as to preserve the numerous fibrous roots which are the principal feeders of the tree. This can only be accomplished by carefully combing them out of the soil with a fork ; and when close to the larger roots of the plant, leaving sufficient soil to preserve them, working carefully under them, transporting them with caution to their destined place on a two-wheeled truck, and, when replanting, spreading out the fibrous roots, and staying the tree. They will not be retarded in growth in the slightest degree. In the Botanic Gardens and the Domain also are ample proofs of the correctness of this theory. Hence it appears that the formal rows of Araucarias, Cunninghami, excelsa, and Bidwilli, Pinus Halipeusis, Cupressus, and many others, which now stretch across that portion of the Botanic Garden towards Government

House, large though they are, can be removed to other parts of the grounds, thus enabling me to make the place upon which they are growing a picturesque and valuable Pinetum, sufficiently large indeed to admit of grouping in their proper order the extensive collection of coniferous plants obtained from Victorian nurserymen during the past year. This spot, on account of its height, would be very appropriate for the purpose; it overlooks the greater portion of the Garden. Many plants of considerable size were lifted from December, 1874, to March, 1875, and placed in suitable spots; all are now flourishing. In all, 832 fine specimens were thus transplanted, and the loss only amounted to about one in every 140, thus dealt with.

The Band Stand still remains on the Palm House Lawn, but I would decidedly reiterate my former opinion, that it should be placed upon the edge of the Lagoon, or on a promontory jutting from it, where the acoustics would be so favorable as to convey the music to all parts of the Garden. The Palm House is in a very bad condition, liable to be blown down by the first heavy gale. The valuable plants it contains afford a reason for remedying this at once. A design (in the preparation of which Mr. S. H. Merrett of the Public Works Department took the principal share in his spare hours) is in the hands of that department; I consider it about the most suitable one to meet the occasion and if carried out would be a source of great attraction to the public as from its size and the way in which it has been planned the large collection of choice tropical plants would be shown to great advantage.

An exceedingly large number of plants have been added to the Botanic Garden collection during the past year. Many of them have been known for years in the nurseries of this city and also in private collections. It was of course advisable to add these to our collection. Others are quite new and have been purchased from various nurserymen. In addition to the plants thus purchased, I have to thank all the Victorian nurserymen for their extreme kindness in affording me every opportunity of selecting plants and seeds gratuitously. A large and valuable collection was thus obtained. I am also greatly indebted to Mr. A. R. Wallis Secretary for Agriculture for valuable assistance rendered me in obtaining plants and seeds from India and America and to Professor Wyville Thomson Chief of the Scientific Staff attached to H. M. S. *Challenger's* expedition. To the latter I am also indebted for a large and valuable collection of seeds and plants from Kerguelen's Land. Mr. D. Sullivan of Moyston also must be specially mentioned for the kindness he has always shown in collecting and seuding the seeds and plants of his district. Mr. McBirnie of the Industrial Museum kindly gave me three cuttings of the shellac plant (Ficus religiosa) which I have been able to multiply. This valuable substance (Shellac) is in India obtained from many plants, and locality favouring production there more in one plant than another, we may yet be able to find some tree in this colony upon which the lac insect would thrive much better than on any of those highly tropical trees. The successful introduction of this insect into the warmer parts of Victoria, New South Wales and the other colonies, would result in the production of a most important article of commerce, and the experiment is well worth trying. I would offer the suggestion to the notice of the Acclimatisation Society. I do not think the attempt has ever yet been made in any of the Australian Colonies. As it is well known that the lac insect does not confine itself to the Ficus religiosa but thrives on many other trees, it is certainly advisable to make the attempt, for if but one species of tree could be discovered suitable to its habits a valuable industry would be created. For instance the State Forests of Victoria might be utilised in a large degree by the introduction of even such an atom as this insect.

The piece of ground known as the Botanic Gardens reserve and intended to be used as a cow paddock for Government House has been added to the Garden proper. It consists of thirty acres and the Garden would never be complete without it. A proper experimental ground upon a *large* scale, for the purpose of growing or acclimatising useful plants for distribution amongst those colonists who would be glad to cultivate them will thus be provided for. Another spot in the Domain can easily be selected for a cow paddock. This annexation will admit of handsome entrances to the Botanic Garden from the South Yarra drive one from Park and Millswyn streets one from Anderson street and another nearly opposite to the botanical Museum near the Observatory. At the latter entrance I have placed "pro tem" the Director's office. The two rooms formerly used for the purpose in my private residence left me insufficient accommodation for domestic purposes. The permanent office however should be either central or at one of the principal entrances easily accessible to the public ; and when practicable it would be highly desirable to add to it a free botanical library and museum of dried plants Carpological specimens &c., for the use of the public.   ,

It is necessary that a fence should be erected along the bank of the Yarra to prevent larrikins from entering the Gardens at forbidden times and when practicable it will be advisable to have lodges built at all the entrances to the Garden.

I have handed in with this year's estimates an item of £200 which I trust will be granted for the erection of two summer houses in the Garden. These are not only required for shelter in case of rain but can be made highly ornamental, being constructed of rustic work, with thatched roofs. A necessary step is the erection of drinking fountains in the grounds. In hot weather it is distressing to visitors to be unable to obtain a drink of pure water. In Fitzroy Gardens the Yan Yean is laid on, but here it has been cut off for some time; and the grounds are supplied by a six horsepower engine which pumps up from the Yarra, water of a decidely inferior quality. I trust that as soon as the new Reservoir now approaching completion, is finished the trouble I have had in being forced to use water carts will be at an end.

With respect to labelling the plants it has always been a difficulty with Botanical Directors, the wear and tear involving continual trouble and expense. Dr. Schomburgk Director of the Adelaide Botanic Garden, has had a system in operation since 1871 which his experience induces him to pronounce successful, and I contemplate employing it here, as it is a well known fact that white labels distributed over the ground have an incongruous and unpleasing appearance. The label should not be perceptible until the visitor closely approaches the plant. To carry out the system effectually, a special vote would be necessary; but once accomplished there would be a considerable saving in future.

The new Catalogue is approaching completion, but my incessant occupation out of doors prevents me from giving that time to it which I would desire, however, as it will be the first of its kind produced here, and as I am adding largely to the contents of the Garden (as will be seen by the annexed list), and assigning common names to almost all plants, in order that non-scientific people may have a plain guide, I wish to make it as perfect as possible. With that object I am devoting to it my spare time after the day's labor is over. From the same cause I have been unable to give that attention to the rockeries I originally intended; the fact being that finding, after repeated trials, the foreman appointed by me so utterly incapable as to be unable to realise the effect ultimately to be produced, I determined—rather than have the intended design spoilt by unskilful hands—personally to superintend the work, at all events until I am enabled to select a man in whom I can place implicit confidence to faithfully carry out my plans. My late foreman during my confinement to the house for some days through severe illness, displayed a glaring want of taste by erecting a pile of stone far more resembling the great wall of China than a natural looking rockery. I was obliged to have the unsightly mass pulled down on my recovery; and even at the present time my hands retain traces of severe bruises received while practically showing how the rocks should be placed. Besides this sort of thing there is a large amount of correspondence to answer daily, from persons writing as to the habits and culture of plants, which I am compelled to attend to. The frequent cases of insubordination which occurred during the past year, point to the fact that unless proper control is given, it is in vain to expect discipline. Any man who considers himself harshly or unjustly treated, has always the power of appeal, but so long as the staff are not immediately responsible to the head of the branch the matter becomes serious, especially where neglect or ignorance may cause the loss of valuable plants, some perhaps scarcely to be replaced. The best way to secure proper service would be to leave with the Director the selection and responsibility of his staff. Such is the rule adopted in all other Botanic Gardens.

The bridge over the lagoon near the Fern Gully should, as mentioned in my last annual report be an arched rustic one faced with rock work, to be in keeping with the surroundings. The fern gully is now one of the attractions of the Garden, and it is gratifying to learn from the remarks of the visitors that the prognostications made at its commencement that it would be a failure have proved incorrect in public opinion. The shade afforded the ferns by the transplanted large trees have greatly aided their development, and the spot is now much frequented by visitors to the Garden.

At the close of this report I have given an amended code of regulations for the safe keeping of the Gardens, which I trust will be adopted. They are almost identical with those now in force in the Botanic Gardens of Sydney Adelaide and Brisbane. Rule 1, for instance, would be a decided improvement. Smoking in the Garden is often a source of annoyance. Larrikins puffing their clouds of smoke into ladies' faces is at present a great nuisance, and one often complained of.

Work requires to be pushed on vigorously in the Domain, to make the construction of the private grounds keep pace with the completion of Government House. Until the crest of the hill fronting the house has been removed by the Public Works Department it will be impossible to form the lawn. This work however is now in progress.

An extra sum of money is needed for other improvements in the Domain. The soil in some places is naturally poor, a large amount of labour is necessary ; lifting large trees to suitable spots necessitates very heavy work ; and the same remark applies to the large quantities of gravel required for walks &c.

When the top of this hill has been removed so that the lawn can be formed and the approaches to the house are complete—which I hope will be the case in the course of six months—I shall then be enabled to make the lawn of buffalo grass, and to arrange the groups and single specimens. There is an abundance of material in the way of plants, specially provided such as they are, propagated from those in the Garden which can be made ready at a moment's notice. In the propagating department, I have endeavoured to multiply all of those ornamental trees and shrubs which will be of use for the Domain and Governor's grounds. Besides these I have saved (by taking up from that portion of the Botanic Gardens near the lagoon previously alluded to where the lawn is being formed) the superfluous shrubs and trees, and planting them in rows ready for use. Nothing has therefore been *destroyed.*

I have ploughed a piece of ground in the Domain for the reception of various grasses, which will be planted on a large scale. I have always borne in mind the importance to the colony generally of this experiment, but want of space in the Botanic Garden prevented me from doing more than propagating the different kinds and planting them in rows until more extended space was available. The Botanic Garden reserve to which I have previously alluded will be of invaluable service in this matter. Amongst these grasses "the Buffalo grass" (of which at the time of mentioning it in my first annual report I had but a small quantity) I have now propagated so largely that I could very shortly be able to supply squatters and others with this splendid grass which besides being of highly ornamental appearance is as I previously stated, a first-class fodder grass tenaciously resisting the most trying heat—a very valuable quality in this colony. I remember that when in Brisbane Mr. Lewis A. Bernays the President of the Acclimatisation Society there, pointed out to me the fact that the deer which were then browsing, selected the little strip of Buffalo grass before any other in the paddock and had nibbled it quite down; and I was often told by gentlemen who had received from Mr. Walter Hill Director of the Brisbane Gardens a few patches to put on their runs that the cattle preferred it so much to other grasses, that the difficulty was to propagate it unless in a specially enclosed paddock, as it was fed down to the roots in summer time. Sheep have been known to eat the roots greedily, and thrive on them, when not a blade of grass was to be had. It must be remembered that though the Buffalo grass as seen upon the edgings I am now making presents a very smooth appearance, that even in poor soils where it is not cut, it attains the height of 18 inches. The broad blades are full of sap and remarkably nutritious. Hence my determination at once to form a collection of grasses in which this one specially valuable alike to the squatter and agriculturalist can be seen attaining its proper height and luxuriance.

The Doub grass (Cynodon dactylon) often erroneously called "doob" is a native of Bermuda ; and in my opinion ranks next to the Buffalo grass (Stenotaphrum glabrum) as a hardy pasture grass for arid climes, though as a lawn grass it is inferior, presenting in the winter a brown and rusty appearance. I can state from experience however in New South Wales and Queensland that where it has been introduced round a station hut, horses and cattle when left to feed as they chose, have collected round it eagerly, refusing the native grasses in luxuriant growth near them so long as a blade of the doub grass remained. As to the nutriment contained in these two grasses, there can be no doubt, as many squatters across the Murray could testify. Respecting their durability the same may be said ; as during seasons of excessive drought, when scarcely a blade of them could be seen, so tenacious of life were they that when the weather broke they sprang up in rich luxuriance ; and when native grasses were totally destroyed by the drought, these two species were the only ones that withstood it. An active interchange of seeds and plants has been kept up with intercolonial and Foreign Botanic Gardens acclimatisation societies, nursery and seed establishments, from all of which large and valuable additions of choice plants and seeds have been received. The vote (£400) placed at my disposal for 1874–5 has enabled me to purchase from Melbourne and other nurseries many plants entirely new to these Gardens. This vote will I trust be supplemented by successive yearly grants, until the world's flora is fairly represented here.

As I cannot possibly visit many parts of Victoria on account of the large amount of work before me, I hope that means may be available for me to send a qualified collector to places where I believe many new and beautiful plants are to be found. For instance, I understand that Wilson's Promontory and many parts of Gippsland have never been botanically explored, and a visit to those parts would be amply repaid by the new specimens discovered.

Considering the importance of the Philadelphian International Exhibition I have prepared so far as my limited means allowed, a collection of fibres, fancy woods various kinds of paper made from native products, gums resins &c. from indigenous trees for exhibition. I trust that though the notice given is short the woods will be sufficiently seasoned to take the polish they are capable of receiving, and which will show them to greater advantage. Should the fibres sent prove of mercantile value the fact is sure to be discovered in such a great gathering of practical men from all parts of the world ; in which case our State forests may supply the materials for local industries and exportation. The collection will be as complete as it can be made. Appended is a list of the principal articles already prepared for exhibition. This colony should certainly be worthily represented in its botanical products as, to say the least of it the exchange of valuable specimens would thus be promoted and attention drawn to the resources of Victoria.

A curious circumstance occurred in connection with the operation of clearing the Lagoon. Three species of weeds had completely covered that sheet of water, viz., *Potamogeton obtusifolius, Heleocharis sphacelata* and *Triglochin procera.* By the aid of a contrivance I devised consisting of two scythes fixed to two pieces of wood T shaped which was towed round the lagoon at the stern of the boat these weeds were cut and removed. Strange to say another weed (one of the confervæ) which had not been seen in quantity before, then made its appearance and is giving infinitely more trouble than its predecessors. Periodical dragging however with the contrivance above mentioned, keeps the lake clear. This weed is as difficult to eradicate as thistles.

Now that a number of spacious lawns are being created in the Botanic Gardens, the provision of additional seats would be a great boon to the public. They might very easily and cheaply be placed round trees in the fashion adopted in the English Public Parks and Gardens. Remarks are frequently made respecting this want of accommodation especially during the hot weather when people become fatigued with rambling about the grounds.

One of the lawns now forming in the Botanic Garden to add to the scenery, I am so creating, as to render it available, should it ever be deemed necessary as an archery and croquet ground for public use. This will not in the least interfere with the intended use of the lawn as an addition to the landscape.

I have the honor to be, Sir,

Your obedient servant,

WILLIAM R. GUILFOYLE,

31st May 1875. Director.

A plan of the Botanic Gardens is herewith appended. The *dark lines* throughout show the condition of the grounds when handed over to my charge on July 1st, 1873; the *red lines and markings* define the landscape improvements effected by me since that date, as well as those which are gradually being carried on towards completion.—W. R. G.

REGULATIONS FOR THE CARE, PROTECTION, AND MANAGEMENT OF THE BOTANICAL AND DOMAIN GARDEN.

WHEREAS by section 108 of *The Land Act* 1869 it is among other things enacted, that the Board of Land and Works shall have power from time to time to make and alter or rescind rules and regulations for the care, protection, and management of all public parks and reserves : Now therefore the said Board of Land and Works, in exercise of the power conferred as aforesaid, doth hereby make the Rules and Regulations following, to be observed and enforced in respect of the Botanical and Domain Garden :—

1. No person shall interfere with the trees, shrubs, flowers, statuary, labels, fountains, fish, or birds therein, or step on the beds, borders, seats, or edges of the grass plots, or engage in any sport or game, or throw stones, or commit any nuisance therein, or leave any bottles, orange peel, paper, cast off clothing, or litter therein, or light fires therein, or smoke, or be in the garden at unauthorized times, or annoy visitors, or convey flowers into the garden, or detain the gardeners by conversation. Visitors shall leave the garden by the nearest path at the time of closing which is notified by the ringing of a bell.

2. No person shall be allowed to climb or jump over the fences therein, or to stick bills on such fences or on the gates, or to cut names, letters, or marks on the trees seats, gates, posts, or fences, or to write thereon.

3. No person shall be allowed to bring any dog into the garden unless led by a chain or cord, or to run any goats or poultry therein.

4. No children under twelve years of age shall be admitted therein unless accompanied by friends or nurses, with whom they shall remain while in the garden.

5. No person shall be allowed to offer for sale any article therein.

6. Unless the Minister of Lands and Agriculture shall otherwise direct, the gates of the Botanic Garden shall be *closed during the summer*, from 6 o'clock p.m. until 8 o'clock a.m., and *in the winter* from 5.30 p.m. to 8 a.m. Any person found taking or injuring plants, flowers or fruits, will be summarily removed from the gardens, or proceeded against by law.

7. Entrance into plots of ground specially enclosed for plantations and for other purposes is prohibited.

8. No person shall be permitted to enter or remain in the garden who is not decorously dressed, and no person shall be allowed to commit any act of indecency either by word or action in this garden. The gardens will not be open to the public on Sundays until after the hour of 1 o'clock p.m.

9. If any person offend against any of these rules such person may be forthwith removed from the garden by a bailiff or constable.

Offenders against these regulations shall in accordance with section 108 of the *Land Act* 1869, on conviction before any justice of the peace, forfeit and pay a penalty not exceeding £5 for each offence ; and every person who shall knowingly and wilfully offend against any such regulation shall be forthwith apprehended by such Crown lands bailiff or constable, and be taken before such justice of the peace, and shall on conviction, forfeit and pay a penalty not exceeding £10.

LIST OF DONORS.

Acclimatisation Society, Melbourne. Several vols. Yearly Reports, &c.

Adelaide Botanic Gardens (Dr. Schomburgk). Miscellancons and select plants.

Adet, Mons., Melbourue. Seeds from New Caledonia.

Agriculture, Department of, Melbonrne. Miscellancons seeds, &c.

Anderson, Hon. R. S., Melbourue. Bulbs, plants, and cuttings.

Archer, W. II., Melbourne. Plants and seeds.

Bacchus, W. H., Peerwur, near Ballarat. Several species native grasses.

Banks, T., South Yarra. Quantity miscellaneous bulbs.

Barlee, Hon. F. P., Colonial Secretary, West Australia. Some very valuable seeds.

Belharry, Mr., Melbourne. A few plants.

Beveridge, P., French Island, Western Port. Plants and seeds.

Bishop of Melbourne (Dr. Perry). Miscellaueous plants.

Bosisto, J., Richmond. A few seeds.

Bowen, His Excellency Sir Geo., Melbonrne. Quantity miscellaucous plants, &c.

Brisbane Acclimatisatiou Society (L. A. Bernays, Esq.). Plants, seeds, &c., in quantities.

Brisbane Botanic Gardens (W. Hill, Esq.). Plants, seeds, &c., in quantities.

Bright, R., South Yarra. Several tree Ferns from New Zealand.

Brussels Botanic Gardens (Mons. L. Lubbers). Miscellaneous seeds.

Brnning, G., St. Kilda. Many large specimen and other plants, in quantities.

Buitenzorg Botanic Gardens (Dr. Scheffer). Collections of plants and seeds.

Bull, W., Chelsea, London. Plants, seeds, &c., in quantities.

Calcntta Botanic Gardens (Dr. G. King). Miscellaneous Indian seeds in quantities.

Cape Town Botanic Gardens (J. McGibbou, Esq.). Collection South African seeds.

Carter, W., Emerald Hill. Several specimen and other plants.

Casey, Hon. J. J., Melbourne. A quantity of valuable seeds.

Casey, N. J., Melbourne. Some seeds from New Zealand.

Ceylon Botanic Gardens (Dr. G. H. Thwaites). Miscellaneons seeds.

Coe, Miss, Emerald Hill. Some seeds.

Cole, J. C., Richmond. Miscellancons plants, bulbs, tubers, &c., in quantities.

Collen, Licut. E. H., Calcutta. Some Indian seeds.

Corbett, E. P., Pietermaritsburgh, Natal. Several large palms, and miscellaucous plants.

Crofts, Miss L., Windsor. A few seeds.

Cummins, Hon. J., Toorak. Bnffalo grass.

Damyon, J., Toorak. New Caledonian plants.

Denton, H. P., Adelaide. Some plants, &c.

Du Bonlay, F., Melbourne. Miscellancons seeds from N. West Australia, and some plants.

Dnncan, W., Malvern. Seeds aud plants.

Edinburgh Botanic Gardens (Professor Balfour). Collections of Miscellancons and select seeds.

Ferguson, Mrs., Immigration Depôt, Melbourne. Some seeds.

Fergnson, W., Macedon. Ferns and other plants.

Ford, R. D., West Melbourne. Miscellaueous flower seeds.

French, C., Botanic Gardens. Plants in quantities, dried specimens, and seeds.

Geelong Botanic Gardens (W. Raddenberry, Esq.). Specimen and other miscellaneous and select plants.

Glenn, C., Fntally, Tasmania. Tasmanian seeds.

Goldie, A., New Zcaland. Miscellaneous plants.

Gordon, T. D., Customs, Melbourne. Victorian seeds.

Greig, J. W., Toorak. Miscellancous plants, cuttings, &c.

Guilfoyle, M., Brisbane. Seeds and plants in quantities.

Guilfoyle, J., Tweed River, New South Wales. Seeds.

Guilfoyle, J. A., Sydney, New South Wales. Seeds.

Gulliver, B., Hobart Town. Quantity of Eucalyptus globulus seed.

Haage and Schmidt, Erfurt, Prussia. Very large and valuable lots of seeds, bulbs, tubers, &c.

Halberstaedter, A., Mt. Brewer, Queensland. Queensland seeds, plants, bulbs, &c., in quantities.

Hannecke, C. F., Rangetiki, New Zcaland. Some New Zealand seeds.

Harding, J., Mt. Vernon, New Zealand. Miscellaneous New Zealand seeds.

Harris, J., South Yarra. Miscellancous plants, in quantities.

Hartmann, C. H., Toowoomba, Queensland. Queensland ferns and other plants, in quantities.

Hayes, M., Melbourne. Some New Zealand seeds.

Henderson, E. G., and Sons, London. Valuable collection of seeds.

Herbert, W., Ballarat. Miscellaneous plants.

Hester, T. J., Prahran. Few seeds from West Australia.

Heyne, E. B., Adelaide. Seeds and plants in quantities.

Hobart Town Botanic Gardens (F. Abbott, Esq.). Quantity of seeds and plants.

Houg Kong Botanic Gardens (C. Ford, Esq.). Select China seeds in quantities.

Hose, Revd. W. C., Tarraville, Gippsland. A choice fern.

Huber and Co., Hyères (Var) France. Valuable lots of seeds, bulbs, tubers, &c.

Independent Church, Trustees of, Collins street, Melbourne. A large and valuable specimen plant.

Ireland, W., Melbourne. Some plants.

Jeffries, J., Geelong. Valuable and select plants.

Johnson, B. and S., Richmond. Select miscellancous plants in quantities.

Johnson, Thomas, Hawthorn. Rose plants.

Kew Royal Botanic Gardens, London (Dr. J. D. Hooker). Valuable and select collections of seeds.

Kilner, F., Rockhampton, Queenslaud. Several collections of Queensland seeds and plants.

Knight, J. G., Palmerston, N. Australia. Some orchids.

Krone, Herr J., German Scientific Expedition. Some miscellaneous ferns and other plants from Auckland Islands.

Lahore, Agri-Horticultural Society of. Collection of Indian seeds.

Lang, T. and Co., Melbourne. Select plants in quantities.

Law, Somner and Co., Melbourne. Fine collection of gladiolus bulbs.

Lucas, R., Colac. Miscellaneous ferns.

Mallett, D., South Yarra. Some select plants.

McMillan, Dr. Ths., Melbourne. Miscellaneous and select seeds.

McMillan, J. R., Richmond. Miscellaneous seeds.

McMillan, T., Prahran. Miscellaneous and select seeds.

McKeuzie, W., Deniliquin, N. S. W. Seeds of native plants.

McBurnie, Mr., Melbourne. Cuttings of Ficus religiosa.

Miller, Hon. H. (per his gardener, Mr. Boyce), Kew, Melhourne. Several fine specimen and other plants.

Miller, F. A. (Miller and Sievers), San Francisco, California. Miscellaneous and choice American seeds.

Milton, J. B., East Melbourne. Quantity of cuttings.

Moran, H., South Yarra. Miscellaneous plants and cuttings.

Mueller, Baron Von, Government Botanist. Several collections of miscellaneous seeds.

Natal Botanic Gardens. Several large palms.

Poolman, F. (Sugar Works), Sandridge. Baskets for plant protection, &c.

Powers, Rutherford and Co., Melbourne. Seeds and specimens of native plants.

Robertson, Mr. (J.P.), Woollan. Some seeds.

Robinson, G. W., Berwick. Quantity of Eucalyptus and Casuarina seeds.

Rockhampton Botanic Gardens (R. S. Edgar, Esq.). Queensland Ferns, and Palm seeds, in quantities.

Sargood, F. T., St. Kilda. Collection Herbaceous plants.

Scott, J. and Sons, Melbourne. Collection miscellaneous and select plants, in quantity.

Snee, Major, Victoria Barracks, Melbourne. Some mosses and ferns.

Soues, E., South Yarra. Tree ferns, and some Botanical specimens.

Spence Bros. and Co., Melbourue. Some South African feru trees.

St. Petersburgh Botanic Gardens (Dr. E. Regel). Several collections of choice seeds, in quantity.

Storck, J. C., Fiji. Collection of valuable plants.

Sullivan, D., Moyston. Seedlings and seeds of native plants, &c., in quantities.

Sydney Botanic Gardens (C. Moore, Esq.). Several large palms, and other miscellaneous and select plants.

Taylor and Sangster, Toorak. Collections of select miscellaneous plants, in quantities.

Thompson, W. K., Melbourne. Plants from Fiji.

Treen, W. H., Melbourne. Some gladiolus bulbs.

Turton, J. S., Fitzroy. Collection of South African plants.

Vettler, J., Echuca. Several species grasses, and some seeds.

Vienna Botanic Gardens (Dr. Fenzl). Collection of seeds, in quantity.

Vilmorin Andrieux and Co., Paris. Very large and valuable lots of seeds, bulbs, tubers, &c.

Waller, W. H., Maldon. Some cuttings of native plants.

Wallin, R., Melbourne. Some seeds.

Wallis, A. R., Vaucluse, Richmond. Plants and seeds.

Watt, D., Richmond. Collections of select and miscellaneous plants, in quantities.

Watters, P., South Yarra. Fern spores.

Webb, W., Prahran. Miscellaneous plants and cuttings.

Wellington Colonial Museum (Dr. Hector). Select New Zealand seeds.

Wharton, G., Hawthorn. Ferns and miscellaneous seeds.

Wilhelmi, C., Dresden. Collection of seeds.

White, W. P. and Co., Melbourne. Quantity of tubers.

Woodd, W. E., St. Kilda. Some seeds.

Wright, Mr., South Yarra. Miscellaneous ferns, and other plants.

Wyatt, C., Frogmore, Geelong. Valuable collections of plants, and some cuttings.

LIST OF PLANTS WHICH HAVE BEEN INTRODUCED INTO THE GARDENS SINCE JUNE 1873.

NEW SPECIES.

Ampelopsis japonica	Anthoxanthum gracile	Arabis alpina
Aloe pulcherrima	Agrostis pulchra	Aralia Veitchii
Abutilon Avicennæ	nebulosa	Arbutus Menziesii
Acacia acuminata	Adenophora latifolia	Argemone hispida
speciosa	Agathæa spathulata	Aster chinensis
odoratissima	Allium Purschii	ledifolius
serissa	Deseglisci	Astragalus canadensis
Catechu	senescens	Azalea Mortii
Albizzia odoratissima	Acer tartaricum	americana
stipulata	Antirrhinum assurgens	Aira cœspitosa
Moluccana	Aristolochia indica	Alopecurus pratensis
Artabotrys intermedia	Asclepias princeps	Avena canadensis
Anosomeles candicans	Asparagus caspius	sempervirens
Aglaonema commutata	Anemone virginiana	Ludwiciana
Acer striatum	Aubretia Græca ·	myriantha
Amorphophallus bulbi-	Abronia umbellata	occidentalis
ferous	Abutilon luteum	planiculmis
Rivieri	van Houttei	Baubinia acuminata
Aristolochia trilobata	Bowenii	purpurea
Alstrœmeria Errenbaulti	Acianthus fornicatus	arborea
peruviana	Anœchtochilus setaceus	malabarica
Antigonum amabile	Ageratum nanum	Brunia nodiflora
Aristea major	Allamanda grandiflora	Banksia Baxteri
Angelouia grandiflora	Hendersoni	Brownii
Alocasia zebrina	violacea	Bouvardia Davidsoui
Alternanthera picta	Alocasia Jenuingsii	candidissima
Anthurium Scherzirianum	metallica	Humboldti corymbi-
Æchmea coraliua	Allium vincale	flora
Arabis alpina	Amaranthus tricolor	Vrielandi
Aquilegia hortensis	Amaryllis vivipara	jasminiflora
jucunda	picta	Bromus brizaeformis
cœrulea	formosissima	Baptisia leucophala
Ageratum cœlestinum	retusa	leucantha
Alyssum Benthami	vittata	Bassia latifolia
saxatile	Amorphophallus cam-	Butea superba
Amaranthus caracasanus	panulatus	Brownea Ariza
hypochondriacus	Anechtochilus petola	coccinea
speciosa	Anemone japonica	grandiceps
bicolor	Anthurium magnificum	Bleischmedia Roxburghiana
bullatus	Aquilegia atrata	Bryonia laciniosa
Areca oleracea	chrysautha	Burtonia scabra
Aralia cochleata	canadensis	Bignonia Roezliana
Asparagus verticillaris	flavescens	Bilbergia Saundersii

Berberis fascicularis
emarginata
orientalis
, tenuifolius
canadensis
provincialis
Ehrembergii
Barringtonia excelsa
samoensis
Bravoa geminiflora
Bromus arvensis
Babiana spathacea
coerulea
purpurea
Bahia lanata
Balsamati grandiflora
Bauhinia variegata
Betula nigra
Bignonia chirere
venusta
Boronia crenulata
serrulata
Bougainvillea brasiliensis
splendens
Bouvardia Humboldti
splendens
Brodiæa congesta
grandiflora
Browallia elata
Brunsvigia toxicaria
Buckinghamia cellcissima
Cacalia aurea
Caladium atropurpureum
atrovirens
amabile
argyrospilum
Brongniartii
esculentum
odoratum
pictum
Calandrina discolor
Calceolaria hybrida
Calendula maritima
Calochortus macrocarpus
Cantua dependens
Cassia florida
pubescens
marylandica

Casuarina Sumatrana
Cephalotus follicularis
Cestrum nocturnum
Chænesthes lanceolata
Chorozema ilicifolium
varium
Cissus amazonica
Lindeni
Citrus decumana
Claytonia superba
Clerodendrou splendens
Thompsonæ
affine
Kæmpferi
Collinsia bicolor
Combretum Princeanum
Comesperma volubilis
Coreopsis palmata
Corethrostylis Schulzeui
Cornus florida
macrophylla
Cosmidium filifolia
Crotou Guilfoylei
cornutum
latifolium
insularis
longifolium
Mortii
medium
variegatum
Schomburgki
oblongifolium
ovatifolium
Crowea saligna
Cuphea Galleottiana
Cyrtodicra chontaleusis
Cytisus Alochingerii
Cocos flexuosa
Centaurea Fenzlii
orientalis
africana
stereophylla
amara
involucrata
adpressa
rubra
americana
Clementii

Centaurea Amberboi
atropurpurea
babylonica
gymnocarpa
Campanula calycanthemus
grandiflora
grandis
latifolia
Trachelium
americana
mexicana
versicolor
peregrina
carpathica
Commelyna stricta
dubia
Karwinskii
clandestina
Kunthiana
Carex pseudo-cyperus
nutans
rosea
Convolvulus althæoides
Gerardi
maritimus
cantabricus
quadricolor
mauritanicus
Cyclotoma platyphylla
Calceolaria scabiosæfolia
Calliopsis Basalis
Cacalia souchifolia
Coreopsis coronata
Colliusia candidissima
multicolor
Crepis barbata
Canna elegans
Callichroa platyglossa
Ceanothus Arnoldii
integerrimus
Cæsalpinia dasyrhachis
ferruginea
coriacea
Caladium metallicum
Costus albescens
Chamæranthemum Bey-
richianum
Cyrtosiphonia sumatrana

Campsidium valdivianum
Callicarpa augustata
Carica aurautiaca
Cattleya Trianae
Clematis rubro-violacea
Chorozema Chandleri
  macrophylla
  Soulangeana
Chironia floribunda
Cordyline amabilis
Cerinthe bicolor
Campanula alliariæfolia
  Vidalli
Capellia biflora
Caryotaxus japonicus
Clematis Jackmanni
  Veitchii
Cupressus Oaklandiana
Calyptrocalyx spicatus
Chamœrops elegans
Cynoglossum officinale
Calandrina umbellata
Coccoloba excoriata
  uvifera
Cratæva Roxburghi
Calophyllum Inophyllum
Calendula superba
Callichroa platyglossa
Cassia sophora
  fistula
  occidentalis
Criuum amabile roseum
Cipodosa sub-scandens
Crotalaria verrucosa
Chamæpeuce diacantha
Chamæranthemum Gaudi-
  chaudi
Clematis Standishi
  Buchani
Caladenia deformis
  barbata
  latifolia
Cryptostylis reniformis
Corysanthus fimbriatus
Chætospora axillaris
Carex inversa
  Grayi
Dahlia arborea

Dahlia imperialis
Dorstenia argentea
Delphinium formosum
Daucus Broteri
Dæmonorops fissus
  salembanicus
  marginatus
Dipterocanthus spectabilis
Dyckia Lemaireana
Dioscorea illustrata
Daphniphyllum Roxburghi
Dorstenia Reidiana
Dianthus nigricans
  gallicus
  siderocaulis
  dentatus
  viscidus
Dieffenbachia lineata
Digitalis aurea
  fulva
  lanata
  tomentosa
Daviesia brevifolia
Dillwynia ericoides
Diplothemium maritimum
Dypsis sp.
Dalechampia Roezliana
Daphne mezcreum
Datura Wrightii
Daubentonia punicea
Dodecatheon Meadia
Dracæna Regina
  atrosanguinea
  Schomburgki
  Boweni
  Chelsoni
  albomarginata
  Robinsoniana
  Hendersoni
  metallica
  Muelleri
  nigrescens
  nobilis
  Verschaffelti
  Wightii
  Youngiana
Diurus sulphurea
  pedunculata

Danthonia bulbosa
  racemosa
Dichelachne crinita
Epacris exserta
  Kinghorni
  Albertus
  paludosa
  Atleana
  acuminata
  Alexandrina
  grandiflora
  corescens
  Copelandi
Erythrina compacta
Eugenia apiculata
Erica cruenta
  ramentacea
  persoluta
  scoparia
  mammosa
  hyemalis
  Mackiana
  alopecuroides
  pyrolæflora
  coccinea
  couferta
  linnæoides
  Giloa
  Hammondi
  colorans
  Loweswitholi
  Andromedæflora
  autumnalis
  Cavendishi
  Tetralix
  vagans
Eryngium dilatatum
Eucalyptus coccifera
  Gunni
  melanophloia
  robusta
Eugenia gracilis
Eupatorium riparium
Elæodendron australe
Elais Guineensis
  melanocacea
Eranthemum secundum
Erythrina corallodendron

Erythrina Bagotensis
Guilfoylei (Parcelli)
Euryeles Amboinensis
Echites melanoleuca
rubro-venosa
Eleusine coracana
Eranthemum strictum
igneum
Elæocarpus Guilfoylei
Eucharis amazonica
Euonymus Veitchii
Ficus cordifolia
syringæfolia
quercifolia
Frenela Endlicheri
Fontanesia phillyræoides
Fourcroya bulbosa
Franciscea latifolia
Lindeni
calycina
uniflora
Fumaria media
Fagus purpurea
Ficus Hardlandi
Livingstoni
salicifolia
Fittonia argyroneura
Pearcei
gigantea
Festuca duriuscula
Billardieri
bromoides
Gladiolus ramosus
gandavensis
purpurea-auratus
Grislea tomentosa
Gymnogramme leptophylla
Gilia achilleæfolia
liniflora
splendens
Gynandropsis pentaphylla
Guazuma tomentosa
Glycine magnifica
Garcinia Livingstoni
Gustavia augusta
Guaiacum officinale
Gypsophila elegans
Gaillardia coccinea

Gaillardia Richardsoni
graudiflora
Gmelina arborea
Gardenia Blumeana
Gamolepis Tagetes
Godetia Lindleyana
Whitneana
Glossodia major
Geodorum dilatatum
Gardenia Shepherdi
radicans
Gazania splendens
Gesnera macrantha
oblongata
Graptophyllum Earli
Guilfoylea monastachys
Glyceria fluitans
Hardenbergia gracilis
Hedypnois cretica
Heppiella uagelioides
Hibiscus coccineus
virginicus
Helichrysum argenteum
Hellenia amarocarpa
Hermannia plicata
Hartwegia comosa
Helminthostachys zey-
lanica
Helichrysum apiculatum
felinum
fruticosum
fulgidum
odoratissimum
Hyophorbe indica
Hura crepitans
Hymenaxis californica
Hemerocalis fragrans
Hovea acutifolia
pulchella
Hydrangea Otaksa
Iris spuria
atomaria
cristata
setosa
chamæris
stilosa
Xiphioides
Isomene calathina

Isomene undulata
Bahieusis
Bona-nox
Huberi
limbata
siberica
Ixia canariensis
hybrida
pallida
Inga Saman
pulcherrima
bigemina
Ixora rugosella
Iberis hesperidiflora
Ipomopsis elegans
Ipomœa Horsfalliæ
tridentata
Ixia amœna
lilacina
rosea
secundo-patens
purpurea
Ixora princeps
rosea
Duffii
Joica bengalensis
Jambosa acida
Kentia gracilis
Kaulfussia amelloides
Kunzea parvifolia
Kniphofia recurvata
Kæmpferia Parishii
robusta
Lettsonia grandiflora
Leca hirta
divaricata
Laurus chloroxylon
Laurea vespertilionis
Lagerstrœmia indica alba
parviflora
Lychuis Haageana
fulgens
Licuala spinosa
Lunaria biennis
Lansbergia Caracassana
Lupinus albo-violaceus
Livistonia subglobosa
Leptospermum parvifolium

Lotus Gebelei
siliquosus
Lyperanthus nigricans
Lantana urticæfolia
Lilium Humboldtii
Libonia penrhoseiusis
Lasiandra macrantha
Leptosiphon roscus
aureus
Lilium Thunbergianum
Linaria modesta
Linum monagynum
Lagurus ovatus
Marauta bella
eximia
Porteana
regalis
sanguinea
tubispatha
Martinezia granatensis
Meuiscium serratum
Mesembryanthemum tri-
palium
Malope malacoides
trifida
Myristica spe
Mirabilis planiflora
Melianthus Trimemanus
Macrozamia zeylanica
Magnolia Nordbertiana
superba
macrophylla
Mathiola bicornis
Mesembryanthemum capi-
tatum
tricolor
pomeridiauum
Mikania Guaco
Meyeuia Vogeliana
Moræa collina
Mirabilis Wrightiana
Milium multiflorum
Nicodemia diversifolia
Nauclea cordifolia
parvifolia
Nemophila discoidalis
Nemesia floribuuda
Narcissus calathinus

Nelumbium Leichardtii
Nepenthes Phyllamphora
Neriue sarniensis
undulata
Nierembergia gracilis
Olea capensis
Oncidium trulla
Ophioglossum pendulum
Œnanthe peucedanifolia
Oreodaphne Californica
Osmanthus ilicifolius
Outea bijuga
Ouviraudra fenistralis
Oxyura chrysanthemoides
Portulaca canariensis
Thelusoni
Gilliesi
Poteutilla arguta
Pothos argyreæa
macrophylla
Primula auricula
cortusoides
Pringlea antiscorbutica
Pultauæa mollis
retusa
rosea
stricta
subumbellata
Pancratium rotatum
Phœnix Leonensis
farinifera
paludosa
Pterostylis concinna
cucullata
Phycella corusca
Phædranassa gloriosa
Pimelia decussata
Primula japonica
Podocarpus araucaroides
Passiflora purpurea
trifasciata
sanguinolenta
Physianthus albens
Pæony Pottsii
Panax fruticosum
Paudanus Veitchii
Samak
Papaver fugax

Pardanthus chiuensis
Pelargonium echinatum
Pedilauthus padifolius
Peperomia argyreæa
Verschaffelti
Peutstemou glaucum
grandiflorum
Phaseolus Ricardianus
Philodendron Lindenianum
Phyteuma virgatum
Pimelia octophylla
Platystemon californicus
Polianthes tuberosa
Poa alpina
Rhodochiton volubile
Rheum officinalis
RhododendronThibaudioides
neillgaricum
Ricinocarpus pinifolius
Rudbeckia hirta .
Spiræa indica
Lindleyana
Scutellaria Ventenati
Silene reticulata
Sabal havanensis
rotundifolia
Saxifraga rotundifolia
Salvia spleudeus
pateus
Sternbergia alba
Stipa elegantissima
Taxus adpressa
Tapis barbata
Tricyrtis hirta
Tigridia canaricusis
Tropæolum minus
canariense
Tagetes lucida
Terminalia Arjuna
tomentosa
Tritonia gracilis
uvaria
maculata
squalida
rosea
Tunica saxifraga
Tectona grandis
Thunbergia chrysops

Thunbergia Hawtayneana
Tecoma fulva
Tetranema mexicanum
Tetratheca verticillata
Trichonema bicolor
  speciosum
Taxicophila Thunbergi
Tacsonia ignea
Tritellia marragana

Tulipa suaveolens
Vitis incarnata
Vallisneria spiralis
Veronica cupressoides
Viburnum Sieboldti
Veronica lanigerum
  Schmidtii
Vernania javanica

Watsonia coccinea
  gigantea
Wollastonia glabrata
Weigela Lemouii
Xanthorrhœa minor
Zephyrauthes chlorolcuca
  tubispatha
  candida

VARIETIES.

Azalea—
  Hanleyana
  Helene Theleman
  Madame de Camaert
    d'Hamale
  Sir H. Havelock
  Model
  Souvenir de Prince
    Albert
  Stella
  Murrayana
  Splendens
  Van Geert
  Columbus
  alba magna
  Son de Pronage
  Amœna
  Columbico
  Duchesse Adelaide de
    Nassau
  Alba
  Bianca
  Bouquet des Roses
  De Wittiana
  Charmer
  President
  Sir Chas. Napier
  coccinea major
  Comte de Flanders
  Standard of Flanders
  Coronata
  Merker
  rosa superba
  Criterion
  mycrophylla
  Magnifica
  Distinction

Azalea—
  Milton ·
  Duc de Massena
  crispiflora
  optima
  Perfection
  President Herman
  Reine des Pasbays
  Juliana
  King Leopold
  Leopold I.
  Duc de Nassau
  Duke of Devonshire
  Comte de Mayence
  Etoile de Gand
  Alex. II.
  Eulalie Vau Geert
  Flag of Truce
  Baronne Viccry
  Fulder's white
  Empress Eugenie
  Gledstauesi formosa
  Duc d'Aremberg
  Glory of Sunning
    Hill
Agave americana var.
  striata
Abutilon Boule de Niege
  Souvenir de Aragon
Achimenes, Masterpiece
Asclepias, purpurascens
  var. tuberculata
  purpurascens var.
  typica.
Ageratum, Imperial Dwarf
  var.
  Wendlandi album

Amaranthus melancholicus
  v. ruber
  hypochondriacus v.
  monstrosus
  hypochondriacus v.
  racemosus
  bicolor v. albicnsis
  caudatus v. luteus
Antirrhinum rupestre v.
  grandiflora
Anosomeles, ovata v. mol-
  lissima
Abutilon luteum v. erectum
  Thompsonae variegata
  Alphouse Karr
  Souvenir de Aragon
  Comtesse Medici Spada
Acalypha marginata v. acu-
  minata
Acer Wagneri laciniatum
Agapanthus umbellatus
  variegatus
Alocasia macrorhiza varie-
  gata
Amaryllis Guilfoylei
  Johusoni
  Lady Parker
Anemone—
  Agnarius
  Agnes
  Belle Emilie
  Bleu Amaible
  Bleu azur
  Commander in Chief
  Duchess de Lotheringen
  Fanny
  flora perfecta

Anemone—
Josephine
Julie
King of the blue
King of the scarlet
La majesteuse
Lady Ardens
Leverrier
L'ornament de l'Nature
Maria Christina
Marie Antoinette
Marie Stuart
Miss Nightingale
Navarino
Parfait
Princess Alice
Reine du Monde
Thalia
L'Eclair
Armida
Aimable Bergere
Bleu fonce
Cramoisine
Coronaria
Cerise tendre
Chapeau rouge
Duchesse d'Albani
Dragoman
Dorinde
Duc de Turenne
Dame d'Honneur
Eclatante superbe
Elegantissima
Euphrosine
Feu de grand Valeur
General Woronzoff
Granville
Hannah More
Imperator
James Watt
Kyanis
Leonidas
La Sultana
La Joyeuse
La Respectable
La Fontaine
L' Etoile du Nord
Martinet

Anemone—
Mulatto
Mon Trésor
Newton
Neptuneus
Proticta
James Defauts
Spectabilis
William I.
Pucelle d'Orleans
Philomela
Prince Jerome
Prince Arthur
Rouge Charmante
Reine Vasthy
Romulus
Rosamunda
Rubro-virens
Rose de Provence
Sir Joseph Paxton
Queen of the Nether-
    lands
Japonica alba
Aphelandra Roezliana
    rosea
Apidistra elatior variegata
Aster Conmista
Haydeni
Aucuba japonica var. ele-
    gantissima
Begonias—
Duchesse
Weltoniensis
Madame de Canville
S. H. Merrett
George Brunning
M. L. J. Ellery
Mrs. Ellery
Gem of Adelaide
Miss Nind
Madame Gruntberger
Alice Anderson
Diadem
juntina argentea
Beech, white variety
Babiana villosa v. purpurea
Bignonia Roezliana villosa
Bouvardia Hogarthi

Bougainvillæa spectablis
    v. glabra
Brunsfelsia americana v.
    latifolia
Brunsvigia hybrida
Buxus sempervirens varie-
    gata aurea
    sempervirens varie-
    gata argentea
Camellia—
Agnes
Albertus
Arch-Duchesse d'Or-
    leans
Arch-Duchesse Au-
    gusta
Azurea
Bealii rosea
Beecan
Bronachia
Cadro
Candidissima
Charlotte
Charlotte Popudoff
Coerulea
Comte de Paris
Cup of Beauty
Czararene
Daviesii
Edith
Eld
Emperor
Eva
Felicia
fimbriata rubra
Flanders
Fridoline
Guilfoylei
Helenor
Henry Favre
Hon. Mrs. Hope
Iris
imbricata
Isabel
Jouvan
Kezia
Lady Belmore
Lady Parker

Camellia—
Lady Hume's blush
Ledia
Leeana superba
Lombardo
Lowi
Madame Paling
Marie Theresa (vera.)
Mathotiana
Miss Emily Manning
Miss Gladstone
Miss Grinley Manning
Miss Knox
Miss Moore
Miss Murray
Miss Mort
Mrs. Berresford
Mrs. Fairfax
Mrs. Day
Mrs. Mort
Monan
Monty
Nicetas
Nilus
ochroleuca
odoratissima
Optima
Plato
Pratti
Preissi
Regia
reticulata
Sebbii
Selina
Souvenir
Sasanqua rosea
Sweetii
Tabbs
Triumphans
Ulpian
Valtavaredo
Vandesia superba
variegata plena
variatissima
Venus de Medici
White Waratah
Woodsii
Wrightii

Camellia—
Xanthus
York and Lancaster
Caladinm—
bicolor splendens
Chantinii
Houletti
Smithii
rubra venium
Wightii
Verschaffelti
De Candolle
Max Kobb
bicolor magnifica
Alphonse Karr
Dr. Lindley
Emperor Napoleon
Schmidtzii
Carnations—
Attila
Prince George
Zonave
Lady Harding
Prince of Wales
Miss Hannaford
Linda
Emma
Amy Robsart
Rosiness
Laurette
Regalia
Dauntless
Ganymede
Purity
Francis (Picotee)
Chrysanthemums—
Ondine
Palmer's Pride
Pink Pearl
Prince Satsuma
Princess Louise
Queen of England
Rifleman
Robert James
Sylvia
The Tycoon
Viscount Thomas
Golden Beverly

Chrysanthemums—
Julie Lagravene
The Damio
King of Anemones
Mrs. Gladstone
Beverly
Golden Christine
White Christine
Viceroy of Egypt
Hereward
Ossian
Princesse Charlotte
Lord Derby
Mr. Murray
Andromeda
Golden Ball
Prince of Wales
Gloria Mundi
Globe white
Firefly
Cedo Nuli (lilac)
aureum multiflorum
Mr. Wyness
Mrs. G. Rundle
Star
Fleur de Marie
Aimeé Ferrier
Annie Salter
Beanté de Nord
Blonde Beauty
Dr. Masters
Empress
Negro
formosum album
Golden Beauty
Golden Standard
Nagasaki violet
Guernsey Nugget
Jardin des Plantes
Lady Godiva
Lady Slade
Lady Talford
Leopard
Marechal Douay
Marquis de Wildermire
Madame Chalamy
Madame de Vatry
Meyerbeer

Chrysthanemums—
  earinatum
  Dunetti
  tricolor
Cineraria hybrida
Cissus alba nitens
Clerodendron Thompsonæ
  variegata
Coleus Guilfoylei
  Display
  Beautiful for ever
  laciniatus
  Verschaffelti
  Wrightii
Colocasia macrorhiza var.
Crataegus sp. (double
  erimson)
Cytisus Laburnum v. quer-
  cifolium
  Laburnum v. atro-
  purpuream
Crinum amabile roseum
Coffea arabica var. (Eden)
Clematis, Lady Bovill
Centaurea Fenzli var.
  moschata v. rubra
  orientalis v. stenolepis
  orientalis multisecta
  angustifolia
Commelina Kunthiana v.
  pallida
Cineraria maritima v.
  candidissima
Cacalia sonchifolia v. lutea
Clitoria ternata v. alba
Campanula carpathica alba
Calliopsis cardamiuifolius
  v. atrosanguineus
Ceanothus Africanus va-
  riegatus
Coprosma Baueriana va-
  riegata
Dianthus—
  chinensis v. hybridus
  chiuensis v. imperialis
  chinensis v. laciniatus
  chinensis v. tartaricus
  moschatum v. scoticus

Dianthus—
  dentatus v. hybridus
  Heddwigii v. diade-
  matus
  Browu's Mule Pink
Delphinium ornatum can-
  delabreum
Digitalis purpurea v. glox-
  iniflora
Dahlias—
  Andrew Dodds
  Attraction
  Bessie
  Bird of Paradise
  Bird of Passage
  Bob Ridley
  Burnes
  Child of Faith
  Coquette
  Corouet
  Crimson Beauty
  Crimson Orange
  Criterion
  Delicata
  Donald Beaton
  Dr. Schwebbes
  Dr. Webb
  Duchesse Malhabe
  Duke of Roxburgh
  Duke of Wellingtou
  Earl Shaftsbury
  Emotion
  Epaulette
  Fair Imogene
  Fanny Sturt
  Fanchouette
  Flambeaux
  Free Boy
  German Daisy
  German Ruby
  German Youth
  Glowworm
  Goldfinder
  Grand Sultau
  Helen Potter
  Hercules
  Hervine
  Hon. Mrs. Trotter

Dahlias—
  Hugh Miller
  Imperial
  John Downie
  Klein Kirkhaauschen
  Lady Darton
  Lady Elcho
  Lady of the Lake
  Lady Peunaut
  Lady Popham
  Lady Paxtou
  Lady Hubert
  Little Mystery
  Little Acorn
  Little Cordula
  Little Daisy
  Little Dear
  Little Julius
  Little Lina
  Little Philip
  Little Prince
  Little Valentine
  Little Singularist
  Little Virginius
  Little Wilhelmine
  Little Wonder
  Leah
  Madge Wildfire
  Marquis of Beaumont
  Miss Mauners Sutton
  Model
  Mrs. Savory
  Mrs. Sophia Elsmer
  Mrs. Turuer
  Mrs. Wyudham
  Mrs. Trotter
  multiflora
  Norah Creina
  Norfolk Hero
  Pauliue
  Peri
  Pluto
  President Lincoln
  Prince of Lilliput
  Prince of Wales
  Prince of Prussia
  Queen Mab
  Remarkable

Dahlias—
Reussenbaby
Rostrilly Jewel
Rostrilly Lad
Shadow
Starlight
Summertide
Sydney Herbert
Unique
Valentine
Vidette
Von Lenipke
White Aster
Annie Keynes
Daphne iudica alba
 odora variegata
Diervilla rosea variegata
Dracaena ferrea rosea
 Gayi
 nigro rubra
Erica persoluta alba
 persoluta rubra
 cerinthoides coronata
 ventricosa erecta
 ventricosa Rothwel-
 liana
 ventricosa rosea
 ventricosa rosea mag-
 nifica
 ventricosa impressa
 ventricosa grandiflora
 ventricosa breviflora
 ventricosa Browni
 ventricosa superba
 ventricosa magnifica
 v. Rollissoni
 cinerea alba
 cinerea purpurea
 cinerea major
 cinerea coccinea
 Rubens
 pellucida v. insulare
Epacris impressa v. rosea
Euonymus aurea margi-
 nata
 japonica v. aurea
 japonica variegata
 latifolius variegatus

Fuchsia—
Arabella
Autocrat
Avalanche
Bacchus
Beacon
Bland's floribunda
Bridal Bouquet
Brigade
Blue Boy
Catherine Parr
Champion
gracilis variegata
Chicago
The Hon. J. Bright
Amphion
Marksman
Lustre
Jolly
Smith's Avalanche
Mrs. E. Bennett
Lewald
Canelle's favorite
Alice
Mandarin
Carnaervon
Silistria
tricolor Beauty
Heligoland
Canary Bird
Little Bobby
Prince of Wales
Ethel
Lady Heytesbury
Commander
Conquest
Constellation
Corsair
Diadem
Dictator
Elfrida
Extraordinary
Favorite of Fortune
Freund J. Durr
Gazelle
Grenadier
Harvest Home
Herald

Fuchsia—
Inimitable
Instigator
Instance
King of the Doubles
Lady Dunbell
Lady Sale
La Favorita
Leah
Lizzie Hexham
Maid of Honor
Majestica
marginata
Marmion
Minnie Banks
Model
Monarch
Mrs. Shirley Hibberd
Norfolk Giant
Octavia
Oracle
Pillar of Gold
Pius the IX.
President
Priam
Princess Alexandra
Purple Prince
Queen of the whites
Rhoderick Dhu
Rose of Denmark
Rustic
S. C. Henchman
Standard
Sunshine
Sunray
The Lord Warden
The Perfect Cure
Tower of London
Treasure
Triumphans
Triumph
Try me oh !
Umpire
Vanqueur de Puebla
Warrior
Warrior Queen
Wave of Life
White Eagle

Fuchsia —	Gladiolus—	Gladiolus—
Planchette	Belle Gabrielle	Lord Byron
Troubadour	Buffon	Le Titien
Funkia spe. variegata	Caffra	La Catherine
Gardenia radicans variegata	Cardinal	Linne
Fortunei variegata	Cornelie	Le Poussin
Gesnera exouiensis	Clemens	Leonard de Vinci
Leopoldii	Cardinal Duval	La Favourite
Godetia reptans v. insignis	Calypso	Marie Antoinette
Gladiolus—	Chas. Dickens	Madame Perriere
Archimedes	Circe	Madame Leslie
Artaban	Cora	Madame Furtado
Aramis	Canary	(Souchet)
Aristotle	Cecil	Madame de Vatry
Cuviu	Chas. Paxton	Meyerbeer
Cnrauti fulgens	Comet	Mirella
Compte de Bresson	Comte de Morny	Milton
Don Juan	Calendulaceum	Montaigne
Edulia	De Candolle	Mazeppa
Gloria mundi	Dr. Schomburgk	Maude
Galatea	Delicata	Madame Basseville
Goliath	Dawn	Madame Sevigne
Hebe	Emile	Mary
Helene	Ellen	Mons. Lebrun d'Albane
Jean d'Arc	Envoy	Meteor
Louis van Houtte	Eleanor Norman	Madame Domage
La Quintine	Eldorado	Madame Vilmorin
Madame Arnoldii	Elata	Napier
Madame Fanny Bouchet	Faith	Newton
Madame Paillet	Fairy	Argus
Madame Conder	Flora	Addison
Mons. Vinchon	Fenelon	Antiope
Prince Imperial	Felicieu David	Adele Souchet
Princess de Montrouge	Fanny Rouget	Adelaide Devale
Surprise	Galileo	Angela
Achille	Herald	Agathe
Annie	Hope	Albans
Anaco	Hindoo	Achille
Attraction	Isabella	Arsinac
Auriga	Jason	Anna
Amina	James Carter	Bijou
Apollon	Jessie	Bein
Ariel	Joan of Arc	Bernard de Jussieu
Antonius	Jas. Veitch	Celine
Beauty	Jenny Lind	Cassine
Bernard de Pallissy	James Watt	Cleopatra
Brenchleyensis	Lord Raglau	Cherubim
Bernice	Lady Bowen	Cuvier

Gladiolus—
Célimène
Didou
De Lamarck
Diomede
Duc de Nassau
Diana
Donna Maria
Erato
Eurydice
Eldorado
Fulton
Greuze
Hortense
Homer
Henri Favre
Ida
Junon
John Bull
Le Dante
Lilacina
Lacipede
Livingstone
Leonora
La Poussin
Largans
Martha
Mirabilis
Marie Dumortier
Mozart
Ninon de l'Enclos
Nestor
Ophir
Oracle
Penelope
Rebecca
Scribe
Sir Jos. Paxton
Thalia
Vulcau
Neptune
Napoleon III.
Novelty
Oscar
Pyrous
Phidias
Princess Mary d'Cambridge

Gladiolus—
Prime Minister
Pallas
Phoebe
Princess Clothilde
Purpurea
Princess of Wales
Pliny
Premier
Prince Fredk. William
Queen of Sheba
Racine
Rana
Roi Leopold
Rossini
Rosenberg
Romulus
Robert Burns
Rev. Berkley
Rajah
R. S. Hill
Rembrandt
Rubens
Reine Victoria
Sulphureus
Silene
Stuart Lowe
Sunshine
Sir W. Hooker
Sol
Susanna
Village Maid
Vesuvius
Vellida
Van Dyck
Victor Emanuel
Venus
Vivian
Valentine
Walter Scott
Walter Hill
W. R. Guilfoyle
Grevillea alpina v. aurea
Hyacinths—
A la' Mode (double white)
A la' Mode (double blue)

Hyacinths—
Alida Catharina
Emilius
Grand Vanqueur
Jaune Suprême
La Virginite
Madame Talleyrand
Maria Theresa
Mars
Othello
Panorama
Pasquin
Regulus
Anna Maria
Albion
Alba maxima
Blocksberg
Baron von Thuyll
Chas. Dickens
Comtesse de la Coste
Corymbosa
Czar Nicholas
Elfrida
Emmeline
Garrick
General Latham
Gloria Florum
Groot Voorst
Howard
Johanna Cornelia
L'Ami du Cœur
L'etincelante
Lord Derby
Lord Grey
Mimosa
Madame Hodson
Mons. de Faesch
Ornament of Nature
Princess Royal
Victoria Alexandrina
Victoria Regia
Hemerocallis fulva variegata
Hydrangea hortensis variegata
Imperatrice Eugenie
Ipomœa quamoclit v. alba limbata v. hybrida venosa variegata

Iris germanica v. atroviolacea

germanica v. suaveolens

Ilex aquifolium aurea marginata

aquifolium v. Silver Queen

aquifolium v. ferox variegatum

aquifolium v. ferox argentea

aquifolium v. albomarginata

aquifolium v. albopictum

Iris Xiphioides v. Alice

Xiphioides v. Atalanthe

Ixia v. Cupid

longifolia alba

purpurea striata

aurantiaca major

Thesius

rosea, variety

Jasmium, Maid of Orleans

Lantana Ne plus Ultra

Leca sambucina v. biserrata

Ligustrum ovatifolium variegatum

Lupinus superbus v. Drummondii

Lilium tigrinum flore-pleno

lancifolium corymbiflorum

lancifolium Harrisonii

Thunbergianum aurea. flore-pleno

Thundergianum grandiflorum

Maximowiczii

Lobelia Paxtoni v. tricolor

Queen Victoria

heterophylla v. major

Mangifera, Bombay variety

Singapore variety

large Malda variety

Gofaulboge variety

Myrtus, double white

Maurandya Barclayana alba

Menziesia polifolia floraalba

Myosotis alpestris floraalba

alpestris rosea

Nephrolepis exaltata v. pilosa

Nerium Oleander alba pleno

Narcissus jouquilla florepleno

Orange trees—

Cluster Orange

Parramatta Orange

Siletta Orange

Kissing Point Orange

Show of Sydney Market

Pentstemons—

Antagonist

Aufidus

Clara

Madame Rendatler

Miss Love

Plantagenet

The Queen

Pelargoniums—

Mrs. F. Abbott

Eugene le Gros

Woman in white

Duchesse

Princess Teck

Pascha

Bertha

Pilot

Archduke

Mrs. Mendale

May Day

Compacta

Princess Helena

Sonnet

Applause

Parisian

Fairy

Ann Page

Clarinda

Prince Hubert

Pelargoniums—

Asteroid

Ruby

Llewellyn

Cynthia

Royal Albert

Magnificent

Alabama

Fanny Gair

Ellen Beck

Ajax

Mrs. Mardell

Cardinal

Perfection

Lady of the Lake

Sunshine

Assembly

Restitution

Rob Roy

Madame Sainton Dolby

Red Cap

A. Gray

Warrior

Marguerite

Harold

Emily

Confident

Envoy

Mrs. Dorling

Jean Sisley

Jeanne d'St. Maur

Lord Derby

Lord Stanley

Leonidas

Madame Mezzard

Miss Depou

Thomas Speed

Unique

Mrs. Pollock

Obedience

Flag of Truce

Canary Bird

Nations' Hope

Sunset

Flower of Spring

Aline Sisley

Crown Prince

Captn. L'Hermit

Pelargoniums—
Imperatrice Eugenie
Madame Racouchout
Madame Rudolph Able
M. C. Glijm
National
Pink Perfection
Signet
Terre Promise
Victor
Victor de Lyon
Wilhelm Pfitzer
Miss Mortimer
Troubadour
Genl. Garibaldi
Alba flora
Miss Dyson
Jerome
Madame Heine
Mary Hoyle
Hebe
Fantastic
Model
Ixion
Bonetta
Mrs. Greig
Fair Ellen
Freedom
Virginie Meillez
Festus
Queen Victoria
Eugene Legereau
Picnic
Frederick
Miss Clendinning
Rosy Gem
Alexandra
Duchess of Sutherland
Mrs. Simpson
Admirable
Monsieur Lierval
Princess Beatrice
Congress
Castanet
Richard Benyon
Bride
Royalty
Princess Mary

Pelargoniums—
Princeps
Hofgaertner Kellerman
John Hoyle
Prince Noir
Oddfellow
Claribel
Heirloom
Gustave Malet
Pansies—
Butterfly
Socrates
Eleanor
Prince
Frenchman
Goldfinder
Hebe
Homer
Jessie Gow
King of the Yellows
Mr. Fred
Pandora
Perennial Phlox—
Boree
Comet
De Bois Duval
Dr. Andry
Dr. La Croix
Flora
Louis Lierval
Madame Rendatler
Neptune
Purity
Surpasse Mdme. Rendatler
Monsieur Robini
Lady Hulse
paniculata alba
Van Houttei
Madame Barrilot
Roi des Roses
Monsieur Rafarin
Sultan
Madame Berniaux
Comtesse de Chambord
Madame Hermine Turenne
Comtesse Duchatel

Perennial Phlox—
Madame Menier
Souvenir
Madame Domage
Monsieur Drourt
George Henderson
Madame Saison
Souvenir de Berrier
Comtesse Mallert
Madame Caillard
Comtesse de Panouxe
Madame Ataris
Moisset
Hugh Low
Princess Alice
Enchantress
Mrs. Bonner
Pæonia lutea variegata
General Bertram
Papaver orientale v. brachyatum
Photinia arbutifotia variegata
serrulata variegata
Pittosporum eugenoides variegata
undulatum variegata
Polianthes tuberosa flore-pleno.
Primula vulgaris, pleno atropurpurea
vulgaris, pleno sulphurea
Ranunculus—
Allan Cunningham
Black Turban
Cerialis
Dr. Horner
General Sylvia
Grand Cæsar
Grandiflora
Monsieur
Princess Sophie
Scarlet Turban
St. Jerome
Vanguard
Victoria
Yellow Turban
White Turban

c

Ranunculus—
Ambassadeur
Arch Duchesse
Adeline
Belle Donna
Belle Maria
Belle Lisette
Bonifaciens
Bonte Held
Comte d'Artois
Charlemagne
Capsicum
Comtesse de Pompadour
Cramoisine
Violet Superbe
Commodore Napier
Comte de Ligne
Venus
Caroline
Doriuda
Eldorado
Ursulla
Epicharus
Felixburg
Gloria florum
Henri Quatre
Habit dectoriale
Toison d'Or
Hasetrubal
Theodora
Kroon van Gent
La Sublime Geisdeliu
La Charmante
La Lingulisére
Lina
La Couronne
L'Enchanteur
Louis d'Or
Ophir d'Or
Orange Picoté
Orphens
Oeillet parfait
Quintinianus
Prosperite
Proserpina
Prince de Orange
Rose d'Espagne
Reine de Violettes

Ranunculus—
Rose de Holland
Silene
aureum punctatum
Lady Clermont
Lord Clyde
Lord John Russell
Lord Palmerston
Michael Waterer
Mirabilis
Mrs. G. W. Heneage
Mrs. Holford
Mrs. John Waterer
Mrs. Thos. Warne
Ochroleuca
Princess Royal
Purity
Rosabel
Schiller
Sir Francis Crossley
Sir Thomas More
Stella
Tippoo Sahib
The Queen
The Warrior
Verschaffeltii
Vulcan
Volcano
Alarm
Auguste van Geert
Chloe
Baron Osy
Caractacus
Charles Bagley
Duchess of Sutherlaud
Everestianum
Everestianum var. ele-
gans
General Cabiari
Giganteum
Grand Arab
Herschel
Iago
Illumination
Jenny Lind
John Spencer
John Waterer
Joseph Whitworth

Reineckia carnea variegata
Roses—
Duc de Rohan
Duchesse de Camba-
ceres
Duchesse de Orleans
Duke of Cambridge
Duke of Edinburgh
Emperor de Maroc
Emperor Napoleon
Eugene Appert
Eugene Boucier
Eveque de Nismes
François Arago
François Premier
General Delaage
Geueral Simpson
General Washington
Gloire de Mousseuses
Grace Archer
Great Western
Hortense Blackett
Jean Goujon
John Hopper
Lady Darling
Lady Manuers Sutton
Lady Robiuson
Louise Van Houtte
Louise Margotten
Madame Boutin
Madame Chas. Wood
Madame Clemence
Joigneaux
Madame Julie Daran
Madame Souppart
Madame Therese Levet
Anna de Diesback
Madamoiselle Anne
Wood
Achille Gounod
Adolph Noblet
Alpaide Rotalier
Baron Gonella
Catherine Guillot
Dr. Jamain
Marechal Neil
Maria Nova
Marie de Bourges

Roses—	Roses—	Tulips—
Marquese Bocella	Souvenir de Montault	Monument
Marquis of Cotellami	Souvenir de Mons.	Thomas Moore
Mathew Mole	Rosseau	Wapen van Leyden
Miss Appleton	Souvenir de Wm. Wood	Yellow Prince
Miss Hepburn	Triomphe de l'Expo-	Admiral Kingsbergen
Maurice Beruadin	sition	Blue Flag
Mons. Woolfield	Triomphe de Lyons	Courone Imperial
Poupre de Tyre	Alphonse Karr	Duke of York
Pierre Notting	Turenne	Marriage de ma Fille
President Lincoln	Vanqueur de Goliath	Purple Crown
President Mas	Taxus Canadensis v. va-	Parrot Tulip
President Willermoz	riegata	Verbenas—
Prince de Rohau	Thujopsis borealis variegata	Royal Duke
Prince Camille de	Tydœa var. Beauty	Snowstorm
Rohan	Tigridia conchiflora v.	Viscaria elegans picta
Princess Alice	grandiflora	Veronica longifolia v. pu-
Princess Beatrice	Thymus citriodorus va-	bescens
Reine de la Cite	riegatus	Vernonia javanica v. ob-
Senateur Vaisse	Tulips—	longata
Sir Chas. Darling	Rex Rubrorum	Watsonia meriana rosea
Sir Henry Manners	Tournesol	meriana excelsa
Sutton	Artis	Weigela rosea v. arborea
Souvenir de Count	Golden Prince	grandiflora
Cavour	Lac Van Rhyu	Zante Currant

COLLECTION OF VEGETABLE PRODUCTS

SELECTED AND PREPARED AT THE BOTANIC GARDENS FOR THE PHILADELPHIA
EXHIBITION, AS PER FOLLOWING LISTS.

List of Wood Specimens, with common names attached; also a short description of their general uses, quality, and geographical distribution.

No. 1. Acacia decurrens : *Willdenow.* Ord. Leguminosæ.—The "Common Wattle." A tree of considerable size. Wood close-grained, hard and tough, extensively used for staves of casks, &c.; takes a good polish. Bark valuable for its tannic properties, and also as a paper-making material. Yields a gum similar to gum arabic. Wood considered one of the best of fuels for heating bakers' ovens. It is of rapid growth, and is found growing abundantly in the colonies of Victoria, New South Wales, Tasmania, and portions of South Australia.

No. 2. Acacia pycnantha : *Bentham.* Ord. Leguminosæ. — The "Golden Wattle." A tree of medium height, and of graceful appearance, especially in the flowering season, when its dense masses of golden blossoms has a beautiful effect in the landscape. Wood dense and close-grained, very tough. Bark extensively used in tanning, also furnishing a valuable paper-making material. Yields a transparent gum, similar to that of A. decurrens. It is of rapid growth, and is distributed throughout the colonies of Victoria and South Australia.

No. 3. Acacia longifolia : *Willdenow*. Ord. Leguminosæ.—The "Long-leaved Wattle." A tall shrub or small bushy tree, of quick growth. Wood takes a fine polish and is beautifully grained. Yields a transparent gum ; bark possibly useful for tanning purposes. It is found in the colonies of Victoria, New South Wales, Queensland, South Australia, and Tasmania.

No. 4. Acacia retinodes : *Schlechtendahl*. Ord. Leguminosæ.—A small-sized tree, wood hard and tough. Yields transparent gum, bark contains tannic properties of considerable value. It is of moderately quick growth, and is found extensively throughout the colonies of Victoria and South Australia in open country and adjacent to water courses.

No. 5. Acacia armata : *Robert Brown*. Ord. Leguminosæ.—The "Prickly Acacia." A tall shrub, growing to a height of twelve feet, extensively used for hedges, for which purpose it is well adapted. Wood very hard and close-grained, useful for manufacture of fancy pipes, rulers, &c., takes a good polish and is very durable. Indigenous to Victoria, New South Wales, South and West Australia.

No. 6. Acacia saligna : *Wendland*. Ord. Leguminosæ. — The "Weeping Acacia." A small tree of drooping habit and quick of growth, suitable for hedge planting. Wood very heavy and tough, but easily worked ; it takes a good polish, and is of a fine grain. Yields transparent gum. Bark no doubt valuable for tanning purposes. Indigenous to Western Australia. Wood specimen grown in Melbourne Botanic Gardens.

No. 7. Araucaria Cunninghami : *Aiton*. Ord. Coniferæ.—The "Moreton Bay Pine." A magnificent tree of pyramidal habit, attains from 150 to 200 feet in height in favourable situations. The wood of this tree is very durable, and is esteemed for common household furniture and other domestic purposes. Yields gum-resin in large quantities, which may yet be of great commercial value. It is of moderately quick growth and very ornamental. It is found growing in various parts of Queensland, and on the banks of the Clarence, Richmond, and Tweed rivers in New South Wales. Wood specimen from tree grown in Melbourne Botanic Gardens, where it has attained a height of forty-eight feet.

No. 8. Araucaria Bidwilli : *Hooker*. Ord. Coniferæ.—The "Bunya Bunya Pine." A magnificent foliaged tree, growing to a height of 150 feet, and having a trunk of great girth. Wood dense, hard, and close-grained ; makes excellent furniture, takes a good polish, and is very durable, commonly known amongst artisans as Queensland pine. Yields gum-resin in large quantities. The seeds are eaten by the aborigines, and are borne in cones nearly as large as a man's head. Indigenous to Queensland. The tree has attained here a height of thirty-five feet. Wood specimen grown in Melbourne Botanic Gardens.

No. 9. Sterculia diversifolia : *G. Don*; syn. Brachychiton populneum : *Robert Brown*. Ord. Sterculiaceæ.—The Victorian "Bottle Tree." One of the "Currijongs" of the aborigines. A glabrous tree, growing to a height of sixty feet, with an enormous trunk, somewhat bottle-shaped in appearance, from which fact it derives its common name. Wood very soft and fibrous. It gives early evidence of decay ; and would no doubt yield a pulp for paper making. The bark, which is in successive layers, is a valuable fibre material, suitable for manufacture of mats, ropes, rough cordage, and paper ; and is very rich in a sweet mucilaginous matter of an agreeable taste. Indigenous to the colonies of Victoria, Queensland, and New South Wales.

No. 10. Sterculia accrifolia : *A. Cunningham;* syn. Brachychiton accrifolium : *F. von Mueller.* Ord. Sterculiaceæ.—The "Flame Tree." A lofty tree of highly ornamental appearance. From the exceeding brilliancy of its flowers it is called The Flame tree by the colonists. It might with great propriety be also called the New South Wales "Lace-bark tree." Wood useful, but of inferior quality, from the facility with which it opens, on account of the fibrous tissues of its structure. The bast furnished by this tree is of the most beautiful lace-like texture, and in my opinion is superior to Cuba bast ; the fibre is suitable for the manufacture of ropes, cordage, mats, &c., and can no doubt be utilised for various other purposes of domestic value ; the refuse of the fibre-yielding material would form no mean substitute for horsehair in stuffing mattresses, saddles, &c., &c. Leaves and young wood rich in mucilage, the pith evidently contains a farinaceous matter. Indigenous to the colony of New South Wales. Wood specimen grown in Melbourne Botanic Gardens.

No. 11. Sterculia fœtida : *Linnæus.* Ord. Sterculiaceæ.—A tall-growing, handsome timber tree. Wood hard, dense, and of a beautiful grain, suitable for furniture, &c. Bark valuable as a fibre and paper material; it possesses tanuic properties also. A native of New South Wales, but found also in the East Indian and Malayan Peninsulas. Wood specimen from branch of tree grown in Melbourne Botanic Gardens.

No. 12. Bursaria spinosa : *Cavanilles.* Ord. Pittosporeæ.—The "Spined or Prickly Box." (This must not be confounded with Eucalyptus melliodora, also called Box.) A shrubby tree, which attains a height of forty feet in favourable localities, although a mere bush in alpine and subalpine situations. Wood extremely hard and durable ; would no doubt make excellent common furniture. It is very suitable (from its rough bark) for rustic work. Found in various forms throughout the Australian continent.

No. 13. Casuarina quadrivalvis : *Labillardière.* Ord. Casuarineæ. — The "Drooping Sheoak." A tree of medium size and of very graceful appearance ; found chiefly along the coast, where it is to be seen growing in sand close to high-water mark ; it is also met with a considerable distance inland. Wood, red, tough, suitable for pick handles, &c., takes a fine polish, and is very durable. Is considered one of the best woods for fuel, foliage valuable as a paper making material. Indigenous to the colonies of Victoria, New South Wales, South Australia, and Tasmania.

No. 14. Casuarina suberosa : *Otto and Dietrich.* Ord. Casuarineæ.—The "Erect Sheoak." A tree attaining a height of forty feet, and yielding wood for fuel and other purposes ; is of a reddish color, and takes a good polish. Found in the colonies of Victoria, New South Wales, Queensland, and Tasmania.

No. 15. Callitris rhomboidea : *Robert Brown.* Ord. Coniferæ.—The "Native Cypress." A shrubby tree, growing to a height of thirty feet. Wood, white and durable. Yields a gum similar to gum sandarac. Indigenous to the colonies of Victoria, New South Wales, South Australia, and Queensland.

No. 16. Acacia melanoxylon : *Robert Brown.* Ord. Leguminosæ.—The "Blackwood," but commonly known amongst the settlers as "Lightwood." A beautiful tree, attaining to a considerable height and girth in favourable situations. Timber, hard and close-grained, but easily worked ; heartwood of a beautiful dark color.

In extensive use for furniture, casks, pannelling, for railway carriages, in the manufacture of various musical instruments, and in turnery. It is very durable, and takes a fine polish. Indigenous to the colonies of Victoria, South Australia, New South Wales, and Tasmania.

No. 17. Callitris rhomboidea : *Robert Brown.* Var. Tasmanica. Ord. Coniferæ· —The "Oyster Bay Pine." This tree grows to a considerable size, and yields good timber and a kind of gum sandarac. Found only in Tasmania.

No. 18. Dammara robusta : *C. Moore.* Ord. Coniferæ.—The "Kauri Pine" of Queensland. A lofty growing timber tree of highly ornamental appearance, attains a height of 150 feet, with stem as straight as an arrow, furnishing excellent spars, planks, &c. Wood light and close-grained, well adapted for ordinary furniture ; takes a good polish and is easily worked. Wood specimen from plant grown Melbourne Botanic Gardens.

No. 19. Duboisia myoporoides : *Robert Brown.* Ord. Scrophularineæ.—The "Cork Wood" of New South Wales and Queensland. A bushy tree, attaining a height of forty feet or more. Wood exceedingly light and soft ; bark very suberose. Wood specimen from plant grown in Melbourne Botanic Gardens. This tree is of very rapid growth.

No. 20. Elæocarpus cyaneus : *Aiton.* Ord. Tiliaceæ.—A shrubby tree, glabrous, and of very ornamental appearance, in some situations attaining a height of fifty and even sixty feet. Wood hard, tough, and close-grained. Found in the colonies of Victoria, New South Wales, and Queensland.

No. 21. Exocarpus cupressiformis : *Labillardière.* Ord. Santalaceæ.—The Native "Cherry-tree." A beautiful tree of cypress-like appearance, growing to a height of thirty feet, with a stem of fifteen to eighteen inches in diameter. Wood of a reddish-brown color, hard, and close-grained, adapted for furniture, and is susceptible of a high polish. Found throughout Australia and Tasmania.

No. 22. Eucalyptus corynocalyx : *F. von Mueller.* Ord. Myrtaceæ.—A small shrubby tree, indigenous to South Australia. Wood specimen from plant grown in Melbourne Botanic Gardens.

No. 23. Eucalyptus occidentalis : *Endlicher.* Ord. Mrytaceæ.—One of the Gums of Western Australia. Sometimes attaining a height of eighty feet. Yields a kind of gum kino. Wood specimen from tree grown in Melbourne Botanic Gardens.

No. 24. Eucalyptus cornuta : *Labillardière.* Ord. Myrtaceæ.—The "Yeit-tree" of Western Australia. A middling size tree of bushy habit. Wood specimen from plant grown in Melbourne Botanic Gardens.

No. 25. Ficus macrophylla : *Desfontaines.* Ord. Urticeæ.—The "Moreton Bay Fig." A large growing, much branched tree, with beautiful dark green glossy foliage. Yields on incision a thick viscid milky juice similar to "Caoutchouc." Wood soft and light, gives early evidence of decay. Found growing on banks of streams in Queensland and New South Wales.

No. 26. Grevillea robusta : *A. Cunningham.* Ord. Proteaceæ.—The "Silky Oak." A magnificent tree, with glossy pinnate foliage, and bearing beautiful orange-colored blossoms. Grows to a height of 100 feet, furnishing excellent timber, of a beautiful texture, prized for coopers' work, &c., takes a fine polish. A thick clammy gum of a pale yellow color exudes from this tree, which is no doubt of commercial value ; bark rich in tannin. It is a native of the colonies of Queensland and New South Wales. Wood specimen from tree grown in Melbourne Botanic Gardens.

No. 27. Hakea aeicularis : *Robert Brown.* Ord. Proteaceæ.—A small tree of bushy habit. Wood hard aud tough. Indigenous to the Colonies of Victoria, New South Wales, and Tasmania.

No. 28. Hakea flexilis : *F. von Mueller.* Ord. Proteaceæ.—A small tree, growing to a height of twenty feet. Wood hard and tough. Found in Victoria and South Australia.

No. 29. Hakea pugioniformis : *Cavanilles.* Ord. Proteaceæ.—A low-growing shrub, suitable for hedges. Indigenous to the colonies of Victoria, New South Wales, and Tasmania.

No. 30. Hakea ulicina : *Robert Brown.* Ord. Proteaceæ.—The "Native Furze." A shrub from eight to ten feet high, makes good hedges. Found in Victoria, New South Wales, and Tasmania.

No. 31. Hakea saligna : *Knight.* Ord. Proteaceæ.—A shrub or small bushy tree, attaining a height of fifteen feet. Wood hard and tough. Indigenous to New South Wales and Queensland. Wood specimen from plant grown in Melbourne Botanic Gardens.

No. 32. Hakea oleifolia : *Robert Brown.* Ord. Proteaceæ.—A small tree, growing to a height of twenty feet ; wood tough, not of any known value. Native of Western Australia. Wood specimen from plant grown in Melbourne Botanic Gardens.

No. 33. Hakea laurina : *Robert Brown.* II. eucalyptoides, *F. von Mueller.*— A beautiful, small growing tree, of drooping habit, bearing remarkable globular, crimson and white colored flowers ; attains a height of thirty feet. Wood tough and heavy. Indigenous to Western Australia. Wood specimen from tree grown in Melbourne Botanic Gardens.

No. 34. Hakea cucullata : *Robert Brown.* Ord. Proteaceæ.—An erect growing, very ornamental foliaged shrub, attaining a height of fifteen feet. Native of Western Australia. Wood specimen from plant grown in Melbourne Botanic Gardens.

No. 35. Hymenanthera Banksii : *F. von Mueller.* Ord. Violarieæ.—A rigid prickly shrub, attaining several feet in height in favourable situations. Wood excessively hard, resembling box in appearance, takes a good polish. Found in Victoria and New South Wales.

No. 36. Leptospermum lævigatum : *F. von Mueller.* Ord. Myrtaceæ.—The "Coast Tea Tree." A tall-growing shrub or small tree, attaining a height of thirty feet. Wood hard and close-grained, very durable when unexposed to atmospheric inflences. Valuable as a hedge plant in exposed situations. It forms dense scrubs on the sea coasts of Victoria, New South Wales, and Tasmania.

No. 37. Lagunaria Patersoni : *Aiton.* Ord. Malvaceæ.—The "Cow-itch-tree" of Norfolk Island. A very handsome tree, attaining a height of thirty feet, and bearing a profusion of beautiful rose-coloured flowers. Wood light, soft, and easily worked. Bark valuable as a fibre and paper material. Found in Queensland and Norfolk Island. Wood specimen from branch of tree growing in Melbourne Botanic Gardens.

No. 38. Melaleuca armillaris : *Smith.* Ord. Myrtaceæ.—One of the native Tea-trees. A shrubby species, growing to a height of thirty feet. It is found on river banks and creeks. Wood hard and dense, takes a good polish, but is not durable if exposed to the weather. Bark suitable for paper making. Indigenous to the colonies of Victoria, New South Wales, and South Australia.

No. 39. Melaleuca decussata : *Robert Brown.* Ord. Myrtaceæ.—A tall-growing shrub, one of the "Tea-trees" of the colonists. This species is found growing chiefly on mountain spurs and ranges in the colonies of Victoria and South Australia.

No. 40. Melaleuca ericifolia : *Smith.* Ord. Myrtaceæ.—The common "Swamp Tea Tree." A large-growing shrub or small tree, sometimes attaining a height of thirty feet. Grows chiefly on banks of rivers and other water courses, and also in swampy places, forming dense scrubs. Wood exceedingly hard when seasoned, used by settlers as rafters for huts, &c., will stand for a number of years if protected from atmospheric influences, but soon decays when exposed. Bark valuable as a paper material. Found in the colonies of Victoria, New South Wales, and Tasmania.

No. 41. Melaleuca uncinata : *Robert Brown.* Ord. Myrtaceæ.—A tall bushy shrub, also one of the native Tea-trees. Wood not of any known value. Indigenous to the colonies of Victoria, New South Wales, South and West Australia.

No. 42. Myoporum insulare : *Robert Brown.* Ord. Myoporineæ.—The "Blueberry tree." A straggling shrub or small much branched tree. Wood white, hard and tough, and, from its rough bark, well adapted for rustic work. An ink or dye could no doubt be expressed from the berries of this tree. It is found, in various forms, throughout Australia and Tasmania.

No. 43. Notelæa ligustrina : *Ventenat.* Ord. Jasmineæ.—The Tasmanian "Iron Wood." A small glabrous tree, growing sometimes to a height of thirty feet. Wood extremely hard, heavy, and close-grained ; extensively used in turnery and ships' tackle ; is of great durability, and takes a good polish. It grows extensively on banks of streams in Tasmania and the subalpine districts of Victoria.

No. 44. Oxylobium callistachys : *Bentham.* Ord. Leguminosæ.—A large-growing bushy shrub, native of Western Australia. Value of the wood at present uuknown. Specimen from plant grown in Melbourne Botanic Garden.

No. 45. Panax sambucifolius : *Sieber.* Ord. Araliaceæ.—The "Elderberry Ash." (This is generally called "Mountain Ash," but I have altered the common name, in order that it may not be confounded with "Eucalyptus Stuartiana," also called Mountain Ash. The latter takes its local name from the character of the wood, the former from its close resemblance to the foliage of the Sambucus, though it resembles somewhat the "Rowan Tree" (Sorbus Aucuparia). A small growing tree, with dark green ornamental foliage and smooth shining bark. Found growing chiefly as an underwood in alpine and subalpine situations. Wood white, close-grained, and tough. Used for axe handles and purposes of a similar nature by wood splitters. Indigenous to the colonies of Victoria, New South Wales, and Tasmania.

No. 46. Pittosporum undulatum : *Ventenat.* Ord. Pittosporeæ.—The "Native Laurel." A beautiful glabrous much branched tree, of shrubby habit, growing in some districts to a height of sixty feet. Wood of a light yellow color, exceedingly hard and close-grained, takes a good polish. Forms beautiful ornamental hedges, and its flowers yield a valuable perfume. It is found on banks of streams in the colonies of Victoria and New South Wales.

No. 47. Pittosporum phillyræoides : *De Candolle.* Ord. Pittosporeæ.—A slender tree, of pendant habit. Wood hard and dense. Found, in various forms, throughout the Australian continent.

No. 48. Pomaderris apetala : *Labillardière.* Ord. Rhamneæ.—The "Native Hazel." In some districts a mere shrub, but in humid forests in subalpine situations a graceful tree, attaining a height of thirty feet or more. Wood white, hard, close-grained, tough, and durable. Indigenous to the colonies of Victoria, New South Wales, South Australia, and Tasmania.

No. 49. Tristania conferta : *Robert Brown.* Ord. Myrtaceæ.—A beautiful glabrous, tall-growing tree. Wood white, dense, close-grained, and durable. Found in New South Wales, Queensland, and Northern Australia.

No. 50. Hakea suaveolens : *Robert Brown.* Ord. Proteaceæ.—The " Sweet-scented Hakea." A beautiful bushy shrub, growing to a height of fifteen feet, suitable for hedges. Wood hard and tough. Wood specimen from branch of plant growing in Melbourne Botanic Gardens. Indigenous to Western Australia.

No. 51. Bedfordia salicina : *De Candolle.* Ord. Compositæ.—The Victorian " Cotton-tree." A tall-growing, remarkable looking shrub or small tree, attaining a height of twenty feet or more ; under side of leaves and young branches covered with a close downy substance like cotton. Wood very heavy and close-grained. Found growing chiefly in mountain gullies and on banks of creeks in Victoria and Tasmania.

No. 52. Acacia juniperina : *Willdenow.* Ord. Leguminosæ.—The "Prickly Wattle." A tall shrub, found growing on the sea coast, where it forms dense scrubs, and also as an underwood in mountain gullies and ranges in Victoria, New South Wales, Queensland, and Tasmania. Wood very hard and tough, esteemed by splitters for maul handles, &c.; bark no doubt rich in tannin.

No. 53. Banksia marginata : *Cavanilles*; syn. B. Australis : *Robert Brown.* Ord. Proteaceæ.—The common " Honeysuckle " of the colonists. A low-growing tree. Wood very heavy, soft, of a reddish color, beautifully grained, susceptible of a high polish. Scattered throughout various parts of Victoria, New South Wales, South Australia, and Tasmania.

No. 54. Hedycarya angustifolia : *A. Cunningham;* syn. H. Cunninghami *Tul.* Ord. Monimiaceæ.—The " Native Mulberry " or " Smooth Holly " of the colonists. A small glabrous tree, of shrubby habit. Found chiefly in fern gullies and ravines adjacent to water in subalpine districts of Victoria and New South Wales. Wood hard and close-grained.

No. 55. Leptospermum lanigerum : *Smith.* Ord. Myrtaceæ.—The "Light or Woolly Tea Tree." A tall bushy shrub or small tree, growing chiefly on banks of water courses in high altitudes in the colonies of Victoria, New South Wales, South Australia, and Tasmania. Wood very dense and heavy, not durable when exposed to atmospheric influences.

No. 56. Olearia argophylla : *F. von Mueller;* syn. Aster argophyllus : *Labillardière.* Ord. Compositæ.—The "Native Musk tree." In favourable situations attaining a height of thirty feet or more. The upper side of leaves are of a deep glossy green ; the under side of a beautiful silvery grey color, slightly tomentose, and emitting a powerful musky odor. It delights in rich humid forest soils, and is chiefly found in fern gullies and ravines in subalpine situations, where it attains its greatest height. Wood highly esteemed in cabinet work and turnery, picture frames, &c. It is beautifully grained, very durable, and equal to, if not superior to, maple. Indigenous to Victoria, New South Wales, and Tasmania.

No. 57. Prostanthera lasianthos : *Labillardière.* Ord Labiatæ.—The "Dog-wood." A shrubby tree, attaining a height of thirty or forty feet in alpine and sub-alpine gullies, but in low-lying districts a mere bush. Wood hard, tough, and close-grained, young saplings suitable for whip handles and fishing rods. Found growing principally in rich damp forest gullies or close to streams, where it forms dense scrubs. It is indigenous to Victoria, New South Wales, and Tasmania.

No. 58. Alsophila Australis : *Robert Brown.* Ord. Filices.—The "Mountain Fern tree."—" Umbrella tree " of the settlers. A beautiful species of tree fern, often attaining 40 feet in height. It is found on the sides and often on tops of high ranges, growing luxuriantly even in exposed situations. The heart of this tree, when stripped of its fibrous coating, planed, and varnished, presents a beautiful and novel appearance, somewhat similar to a South Sea Islander's carved war club.

No. 59. Dicksonia antarctica : *Labillardière.* Ord. Filices.—The common Fern-tree of the gullies. Unlike its neighbour, the Alsophila, this ferntree is found only in the most densely sheltered gullies, where in some instances the daylight scarcely penetrates. It attains a great height in rich soils, and presents a magnificent appearance. The heartwood is treated in a similar manner to that of the Alsophila.

No. 60. Cassinia aculeata : *Robert Brown.* Ord. Compositae.—A white flowering bushy shrub, growing to a height of twelve feet or more. Wood hard, not of any known value. Indigenous to Victoria, New South Wales, and Tasmania.

No. 61. Atherosperma moschata : *Labillardière.* Ord. Monimiaceæ.—The "Victorian Sassafras " tree. In the Dandenong and Yarra Yarra range this tree grows to a great size ; its leaves are glabrous, of a light green on the upper side, and glaucous underneath. Wood hard and close-grained. All parts of the tree emit a very strong aromatic odor. Bark valuable for its astringent properties. Indigenous to Victoria and Tasmania.

No. 62. Myrsine variabilis : *Robert Brown.* Ord. Mrysineæ.—The "Smooth Beech " of some districts. A small glabrous tree of shrubby habit, but attaining a considerable height in humid forests. Leaves of a dark glossy green color. Wood very hard and durable. Indigenous to the colonies of Victoria, New South Wales, and Queensland.

No. 63. Melaleuca squarrosa : *Smith.* Ord. Myrtaceæ.—The Victorian "Yellow Wood." In some localities a mere shrub, but in parts of Gippsland a tree attaining a height of 30 feet or more. Wood hard, dense, and close-grained, of a light yellow color. Like many other species of this genus will last a long time if not exposed to the weather. It is found inhabiting marshy places, banks of rivers and creeks in the colonies of Victoria, New South Wales, South Australia, and Tasmania.

No. 64. Lomatia Fraserii : *Robert Brown.* Ord. Proteaceæ.—The "Native Holly." A tall shrub or small tree, sometimes attaining a height of sixty feet, Leaves glabrous, varying in length from 6 to 8 inches, and having deep irregular serratures. Found growing chiefly in mountain regions and deep forest glens, as an underwood, in the colonies of Victoria and New South Wales.

No. 65. Pittosporum bicolor : *Hooker.* Ord. Pittosporeæ.—The Victorian "White-wood." A graceful tree, inhabiting moist fern gullies and ravines, where it attains, sometimes, to a height of fifty feet. Indigenous to the colonies of Victoria and Tasmania.

No. 66. Coprosma hirtella : *Labillardière.* Ord. Rubiaceæ.—The "Native Woodbine." A glabrous, erect shrub. Found growing luxuriantly, as an underwood, in rich, damp forests, in upland situations, where its bright scarlet berries form a fine contrast to the dense foliage with which they are surrounded. Indigenous to Victoria, New South Wales, and Tasmania.

No. 67. Acacia dealbata : *Link.* Ord. Leguminosæ.—The "Silver Wattle" of the colonists. A beautiful tree, attaining to a height of 100 feet in favourable situations. Foliage excessively glaucous, closely resembling that of A. decurrens in form. Wood hard, heavy, and close-grained, very durable : esteemed for casks, and is susceptible of a high polish. The tree is found inhabiting banks of rivers and creeks, and is of exceedingly rapid growth. Its bark is most valuable for tanning purposes, and might also be converted into paper. Indigenous to the colonies of Victoria, New South Wales, and Tasmania.

No. 68. Fagus Cunninghamii : *Hooker.* Ord. Cupuliferæ.—The "Native Beech." A tree of gigantic proportions, attaining a height of 220 feet or more, with a girth of stem of from 40 to 45 feet. It is found growing luxuriantly on the Yarra Yarra Ranges and other districts of Victoria, and also in Tasmania, where it is known to the colonists as the "Myrtle-tree."

No. 69. Dicksonia squarrosa : *Swartz.* Ord. Filices.—A slender tree-fern, growing to a height of twenty feet in moist, rich soils. Generally supposed to be indigenous to New Zealand only, but found also by Mr. D. Boyle in the Dandenoug Ranges. The heartwood is treated in a similar manner to that of Dicksonia antarctica (No. 59).

No. 70. Leucopogon Richei : *Robert Brown.* Ord. Epacrideæ.—"The Currant-wood." A rigid, straggling shrub or small tree, growing to a height of four to fifteen feet. Wood of a reddish color, hard, dense, close-grained, exceedingly heavy and very durable ; it takes a good polish, and from its rough bark and tortuous habit, it is well adapted for rustic work. The berries of the plant somewhat resemble currants in size and appearance, and are edible. It is found growing, extensively along the coasts of Victoria and also on those of New South Wales, Queensland, South and West Australia, Tasmania, and Chatham Island.

No. 71. Banksia integrifolia : *Linnæus, fil.* Ord. Proteaceæ. —"The Coast Honeysuckle." This species is found extensively along the coast, in barren sandy wastes, sometimes assuming the character of a large tree, at others that of a low shrub. In the vicinity of Mordialloc and Frankston, and on Phillip Island (Victoria), trees forty feet or more in height, and having a girth of stem of six to nine feet, are frequently met with. The tree is of very ornamental appearance, when well grown. The under side of the leaves are covered with a short, white tomentum, and when agitated by the wind flash out like silver, presenting a pleasing spectacle, especially in the twilight. The wood is soft, beautifully grained and is susceptible of a high polish. The plant is also indigenous to Queensland and New South Wales.

No. 72. Acacia longifolia, variety sophoræ : *Robert Brown.* Ord. Leguminosæ. —"The Coast Acacia." Usually a decumbent shrub, but sometimes attaining a height of twelve feet or more. It is found growing extensively on barren sandy, and rocky places along the coasts of Victoria, New South Wales, Queensland, South Australia, and Tasmania. Wood fissile, close-grained and hard, takes a good polish and is very durable.

No. 73. Acacia verticillata : *Willdenow.* Ord. Leguminosæ.—A tall shrub or small tree growing to a height of fifteen or twenty feet. Wood, close-grained, hard and tough, takes a good polish and is durable. Found chiefly in humid mountain districts of Victoria. It is also a native of Tasmania.

No. 74. Helichrysum cinereum : *F. von Mueller.* Ord. Compositæ.—A tall-growing, straggling shrub or small tree, sometimes attaining a height of twenty feet. Found along the coast and for a considerable distance inland. Wood, white and hard. Indigenous to Victoria, New South Wales, and Tasmania.

No. 75. Acacia longifolia, variety, mucronata : *Willdenow*. Ord. Leguminosæ. —A tall shrub or small tree. Wood, hard and close-grained, takes a good polish and is durable. Indigenous to Victoria and Tasmania.

No. 76. Acacia stricta : *Willdenow*. Ord. Leguminosœ.—An erect shrub, from three to twelve feet in height. Wood very hard and tough. Found growing in Victoria chiefly near the seaboard. Also in the colonies of New South Wales and Tasmania.

No. 77. Avicennia officinalis : *Linnæus*. Ord. Verbenaceæ.—"The Native Mangrove." A glabrous shrub or small tree, sometimes growing to a height of twenty feet. The shores of Western Port Bay (Victoria), are densely clothed with this plant. It is found chiefly on mud flats, where it grows luxuriantly down to low-water mark, and at flood tide presenting not an unpleasant picture. It is found extensively along the sea coast of most parts of the Australian continent, New Zealand, Tropical Asia, Africa, and America. The wood is exceedingly tough, and is used for mallets, &c. It is very durable under water and for underground work and foundations, but when exposed to atmospheric influences, like the Teatree, it soon perishes. The ashes from this wood supplies one of the best kinds of potash. Its bark is rich in tannin.

No. 78. Melaleuca Preissiana ? *Schauer*. Ord. Myrtaceœ.—"The Mountain Teatree or Ironwood of Phillip Island." A handsome tree. Unlike most melaleucas this species, has a rough bark. On Phillip Island, this tree may be seen, perhaps, to its greatest advantage growing on the beach, where it attains a height of 20 to 35 feet with a diameter of 1 to 2 feet. Wood, very heavy, white, close-grained, exceedingly hard and tough, it is very durable, takes a good polish and is used for many domestic purposes. The tree is indigenous to the colonies of Victoria, South and West Australia where it is found chiefly along the sea coast.

No. 79. Casuarina distyla : *Ventenat*. Ord. Casuarineæ.—"The Stunted Oak." A low-growing, rigid shrub, seldom exceeding 10 feet in height. It is extensively distributed through the colonies of Victoria, New South Wales, South and West Australia, and Tasmania, chiefly on heath grounds and marshy places. Wood close-grained and tough, suitable for handles of tools.

No. 80. Dodonæa viscosa, variety conferta : *G. Don*.—"The Victorian Lignum Vitæ." A tall-growing shrub or slender tree. Found growing principally on barren sandy ridges, close to the sea-shore ; and in some places forming, with Leptospermum lævigatum, dense inaccessible scrubs. The wood is very close-grained and heavy and of extraordinary hardness and durability. The heartwood is of a greenish black color, and is suitable for sheaves for ships' blocks, rulers, tree-nails and fancy cabinet work.

No. 81. Syncarpia laurifolia : *Tenore*. Ord. Myrtaceæ.—"The New South Wales Turpentine-tree." A tall slender tree, of graceful appearance. Indigenous to New South Wales and Queensland. Wood, hard, tough and durable. The tree furnishes a valuable resin. Specimen from branch of tree growing in Melbourne Botanic Gardens.

No. 82. Leptospermum scoparium : *Forster*. Ord. Myrtaceæ.—A tall, bushy shrub in favourable situations, but in alpine districts low and prostrate. Wood, close-grained, hard, durable, and nicely shaded ; takes a good polish. Indigenous to the colonies of Victoria, New South Wales, South Australia, Queensland, Tasmania, and New Zealand.

List of Barks, &c., from which Paper and Fibre for various purposes have been prepared at the Melbourne Botanic Gardens.

No. 1. Paper prepared from the bark of the "Paper Mulberrry" tree.— Broussonetia papyrifera. The uses for which this bark is employed, are too well known to be worthy of remark here. The sample of paper exhibited, in a rough state, has been obtained from a plant grown in the Melbourne Botanic Gardens.

No. 2. Paper prepared from Salvia canariensis.—This plant, a native of the Canary Islands, has been acclimatised in the Gardens, where it grows very profusely and can be easily propagated. Its fibre producing properties are very great, and no doubt of considerable importance, as it furnishes material for paper of superior quality for writing, packing, &c. Sample prepared in rough state at Melbourne Botanic Gardens.

No. 3. Paper prepared from the bark of Dais cotinifolia, a beautiful tree belonging to the order Daphnaceæ; native of South Africa, and growing to a height of twenty-five feet or more. The bark, which peels readily, yields a good material for paper of a fine texture and very white, and also a yellow dye. Sample prepared from plant growing in Melbourne Botanic Gardens.

No. 4. Paper also fibre prepared from bark of the "Grass-cloth" tree of Queensland. Pipturus propinquus : *Wedd.* Ord. Urticeæ.—This tree attains a height of fifty feet in favourable situations, furnishing large quantities of fibre yielding bark suitable for paper of good quality, and also (in a young state) for the manufacture of ropes, fishing nets, &c. It is also rich in tannin, and yields a valuable dye. It is a native of Queensland and New South Wales, but is found also scattered throughout the South Sea Islands and Indian Archipelago Sample prepared in a rough state from plant growing in Melbourne Botanic Gardens.

No. 5. Fibre and bast (in various stages) prepared from the bark of the "Flame-tree" of New South Wales. Sterculia acerifolia : *A. Cunningham.* Ord. Sterculiaceæ.—The bast furnished by this tree is of the finest lace-like texture possible, and is no doubt superior in many respects to Cuba bast. On large trees the bark is fully two inches in thickness. The fibre can be prepared very simply by a steeping process, and is suitable for the manufacture of ropes, strong cordage, mats, baskets, fishing nets and lines, and paper of superior quality. The refuse, which is of a very elastic nature, could be used for stuffing mattresses, saddles, &c., &c. Samples prepared at Melbourne Botanic Gardens from plants growing there.

No. 6. Fibre and bast prepared from the bark of the Victorian "Bottle-tree" (Currijong of the aborigines). Sterculia diversifolia : *G. Don.* Ord. Sterculiaceæ.—The bast furnished, in large quantities, by this tree is somewhat like that of Sterculia acerifolia, but much coarser and of a darker color. It is suitable for ropes, coarse cordage, matting, baskets, paper, &c., and can be prepared very simply by steeping. This tree is found in the colonies of Queensland and New South Wales, as well as in Victoria. Samples prepared at Melbourne Botanic Gardens.

No. 7. Fibre prepared from the bark of Sterculia fœtida : *Linnæus.* Ord. Sterculiaceæ.—This tree, which is a native of the East Indian and Malayan Peninsulas, as well as of New South Wales, furnishes a valuable bark for papermaking, coarse ropes, bags, matting, &c. Its preparation is very simple. Sample prepared at Melbourne Botanic Gardens from tree growing there.

No. 8. Fibre prepared from the bark of Abutilon venosum : One of the "Lantern flower" trees. Ord. Malvaceæ.—A native of Brazil. Fibre of a fine texture suitable for whipcord, fishing lines and textile fabrics, also for paper. Sample prepared from plant grown at Melbourne Botanic Gardens.

No. 9. Fibre also paper prepared from bark of the "Cow-itch-tree" of Norfolk Island. Lagunaria Patersonii : Aiton. Ord. Malvaceæ.—Fine, strong and glossy suitable for paper, of superior quality, ropes, strong cordage, fine matting and basket work. Indigenous to Queensland and Norfolk Island. Sample prepared from plant growing in Melbourne Botanic Gardens.

No. 10. Samples of fibre prepared in various ways from bark, also paper prepared from the leaves of the Chinese "Grass-cloth" plant. Bœhmeria nivea.—This plant is of rapid growth and attaius to great perfection in Victoria. Samples prepared at Melbourne Botanic Gardens.

No. 11. Samples of fibre prepared from the green and also from the dead bark of Sparmannia Africana.—Though a native of South Africa this plant is of very quick growth in Victoria, where it attains a height of at least ten feet. The fibre, which is produced in very large quantities, is of the finest silky texture and of a beautiful silvery white color; it is very easily prepared and is suitable for textile fabrics. The sample prepared from the dead bark is also very strong and would no doubt make good ropes, cordage, &c. This plant will produce two crops of canes in a season, and in my opinion is equal if not superior to the Chinese grass-cloth plant as a fibre producing material. Samples prepared at Melbourne Botanic Gardens.

No. 12. Fibre prepared from the bark of Laportea Gigas : Wedd. Ord. Urticeac.—The "Tree Nettle" of Queensland and New South Wales, where it attains a height of from 80 to a 100 feet. The wood is soft and fibrous and might be pulped-up for paper. The bark furnishes a very strong and fine fibre suitable for whipcord, fishing lines, &c. The natives avail themselves of this bark for fishing lines and nets, but the fibre obtained from the roots is most prized by them for this purpose. Sample prepared from plant growing in Melbourne Botanic Gardens.

No. 13. Fibre prepared from bark of Sida retusa : Linnæus. Ord. Malvaceæ.—The "Queensland Hemp." This plant has established itself at Melbourne and has become very plentiful in the Botanic Gardens, where the samples have been prepared. The bark is suitable for fine paper twine, &c. It is of quick growth in Victoria and seeds very freely by which means it is easily propagated.

No. 14. Samples of fibre prepared from both the green and dead leaves of the "Spear Lily" of East Australia. Doryanthes excelsa : Correa de Serra. Ord. Amaryllideæ.—This plant is of moderately quick growth in Victoria. Its leaves are oue mass of fibre of great strength, suitable for strong ropes, cordage, mats, baskets, brushes, &c.; also a good paper material. The plant somewhat resembles the Fourcroya gigantea of South America in habit and appearance. Samples prepared from plants growing in Melbourne Botanic Gardens.

No. 15. Paper made from leaves of Marica Northiana : Ord. Irideæ.—This plant, a native of Brazil, thrives well in Victoria and furnishes a valuable paper of good texture and of a fine rich yellow color. It is easily propagated by division of the roots. Sample prepared at Melbourne Botanic Gardens.

No. 16. Paper prepared from the stems and leaves of Scirpus fluveatilis : A. Gray. Ord. Cyperaceæ.—A species of "club rush" found growing plentifully on the banks of streams and lagoons in Victoria and other parts of Australia. Yields large quantities of valuable material suitable for writing, printing and packing paper. It is gregarious in its habit and can be gathered with great facility. Samples prepared at Melbourne Botanic Garden.

No. 17. Paper prepared from Typha augustifolia : *Linnæus.* The "Native Bulrush."—This plant, which is available in large quantities in many parts of Victoria and the other Australian colonies, furnishes a first-class paper material for packing purposes, and might, with proper machinery, be converted into a good writing paper. Samples of bleached and unbleached paper in a rough state prepared at Melbourne Botanic Garden.

No. 18. Samples of fibre prepared from the common "New Zealand Flax." Phormium tenax : *Forster.* Ord. Liliaceæ.—The material furnished by this plant is now so well known that it needs no comment here. The samples are sent merely to illustrate to what perfection this valuable plant attains in Victoria, where it is of very quick growth and is easily propagated. A sample of paper is also sent prepared from this plant. Exhibits grown and prepared at Melbourne Botanic Gardens.

No. 19. Fibre prepared from the bark of Abutilon mollis.—Although a native of South America this plant is of exceedingly rapid growth in Victoria and seems to have thoroughly acclimatised itself. Its fibre is very strong and suitable for ropes, cordage matting, baskets, &c., also a good paper material. Sample prepared at Melbourne Botanic Garden.

No. 20. Fibre also paper prepared from the leaves of Dianella latifolia : syn. D. Tasmanica: *Hooker.* Ord. Liliaceæ.—This plant delights to grow on the banks of creeks, &c., where its leaves attain a length of from two to six feet. Yields fibre in large quantities, suitable for mats, baskets, ropes, cordage and paper. It is found in Victoria, principally in high altitudes, and in many parts of Tasmania. Samples prepared at Melbourne Botanic Garden.

No. 21. Fibre prepared from the stems of Caryota urcus. The "Jaggery Palm" of India, which is found also on the north-east coast of Australia. Sample from plant grown in Melbourne Botanic Gardens.

No. 22. Fibre prepared from the leaves of Cordyline indivisa.—The tall-growing "Palm Lily" of New Zealand. All parts of this tree are composed of a fibrous substance, especially the leaves, which yield a very strong elastic fibre, in almost incredible quantities, suitable for the manufacture of strong ropes, cordage nets, mats, paper, &c. The stem can also be utilised for various purposes. This plant is of very quick growth in Victoria, and produces abundance of seed from which it is easily multiplied. The preparation of the fibre is very simple. Samples prepared at Melbourne Botanic Garden.

No. 23. Fibre prepared from the leaves of Fourcroya gigantea.—"The Giant Lily" of South America. This plant attains to great perfection in Victoria and is of moderately quick growth. Sample of fibre prepared at Melbourne Botanic Gardens.

No. 24. Fibre prepared from the "American Aloe," Agave Americana : *Linnæus.* Ord. Amaryllideæ.—Sample prepared from plant grown at Melbourne Botanic Gardens.

No. 25. Fibre prepared from the bark of Dombeya Natalensis : *Sonnerat.* Ord. Sterculiaceæ.—A most beautiful flowering shrub, or small tree, native of Natal. It is of very quick growth in Victoria. Fibre suitable for paper making, ropes, cordage, &c. Sample prepared at Melbourne Botanic Gardens.

No. 26. Fibre prepared from the leaves of Yucca gloriosa: *Willdenow.* ("Adam's Needle.")   Ord. Liliaceæ.—A native of America, but growing to great perfection in Victoria. Its leaves are very rich in fibre of good texture, suitable for ropes, cordage, and matting, and would no doubt yield a strong packing and writing paper. Sample prepared at Melbourne Botanic Gardens.

No. 27. Paper prepared from Carex appressa: *Robert Brown.* Ord. Cyperaceæ. —A species of sedge grass which grows very plentifully on the margins of rivers, creeks lagoons, &c., throughout Victoria. Yields a valuable pulp for paper of a strong coarse texture, but with proper appliances a good writing paper might be obtained. Sample prepared at Melbourne Botanic Gardens.

No. 28. Paper prepared from the stems of Isolepis nodosa : *Robert Brown.* Ord. Juneaginæ.—A rush found growing plentifully on river banks, and marshy places and yielding a valuable material suitable for packing and writing paper.

No. 29. Paper prepared from the stems of Juncus maritimus : *Lambert.* Ord. Juneaginæ.—The "Sea coast Rush." Found growing extensively along the coast and in salt marshes throughout Australia. Sample prepared at Melbourne Botanic Gardens.

No. 30. Paper and fibre prepared from the stems of (small form) Juncus vaginatus : *Robert Brown.* Ord. Juneaginæ.—"The Sheathed Rush." Found growing very plentifully throughout Australia. Yields a valuable pulp for paper making, and can be collected in large masses with ease. Samples prepared at Melbourne Botanic Gardens.

No. 31. Fibre prepared from bark of the Queensland "Bottle-tree." Sterculia rupestris : *Bentham.* Syn. Brachychiton Delabechii : *F. Mueller.*—This tree attains a considerable height and has an enormous bottle-shaped trunk. Its bark is very thick, and yields a very strong fibre, suitable for manufacture of rope, strong cordage, matting, paper, &c. Indigenous to Queensland. Fibre prepared from tree growing in Melbourne Botanic Gardens.

No. 32. Paper prepared from "The Native Nettle." Urtica incisa : *Poiret.* Ord. Urticeæ.—This plant yields a pulp of very fine texture, which becomes a beautiful white color when bleached, producing what seems to be a very valuable paper. Sample prepared at Melbourne Botanic Gardens.

No. 33. Fibre prepared from bark of Commersonia Fraseri : *J. Gay.* Ord. Sterculiaceæ.—A tall-growing shrub or small tree. Found growing chiefly on the banks of rivers in the colonies of Victoria, New South Wales and Queensland. Its bark is extensively used by the settlers as a tying material and yields a fine fibre, in large quantities, suitable for rope, cordage, paper, &c.

No. 34. Fibre prepared from bark of Abutilon Bedfordianum : *Hooker.* Ord. Malvaceæ.—One of the "Lantern flowers." A tall rank growing shrub, native of Brazil. The bark of this shrub, which grows quickly in Victoria, yields a fibre of a very superior order, suitable for whipcords, fine matting, paper, and, perhaps, textile fabrics. Sample prepared from plant grown in Melbourne Botanic Gardens.

35. Fibre prepared from bark of Plagianthus pulchellus: *A. Gray.* Syn. Sida pulchella : *Bonpland.*—A small shrubby tree, found growing extensively on the banks of the Yarra Yarra and other rivers in Victoria, also in the colonies of New South Wales and Tasmania. It yields a very fine bast, the fibre of which is very strong, and suitable for manufacture of whipcord, fishing lines, nets, fine matting, and paper. Sample prepared at Melbourne Botanic Gardens.

No. 36. Fibre prepared from dead leaves of the "Screw Pine" Pandanus utilis : *Bojer*. Ord. Pandanaeceæ.—Sample prepared from plant growing in Melbourne Botanic Gardens. Indigenous to Mauritius.

No. 37. Fibre prepared from bark of Abutilon striatum : *Dickson*.—The striped "Lantern flower." This shrub is of exceedingly quick growth, it is a native of Brazil, but thrives remarkably well in Victoria. The bark, which peels readily, furnishes a fibre of a very fine texture, which is very easily prepared. This plant might be grown with advantage as a fibre yielding material ; two crops of canes of considerable length might be obtained in a season under favourable circumstances. Sample prepared at Melbourne Botanic Gardens.

No. 38. Paper prepared from Ehrharta tenacissima : *Nesb*. Ord. Gramineæ.— A tall growing wiry grass, which delights to insinuate itself amongst the lower branches of shrubs and trees. Can be obtained in large quantities in the upland regions of Victoria and other parts of Australia. Furnishes pulp suitable for packing and writing paper. Sample prepared at Melbourne Botanic Gardens.

No. 39. Fibre and paper prepared from the bark of Pimelia axiflora : *F. von Mueller*.—"Currijong" of the Aborigines.—A tall growing glabrous shrub, with a smooth brown bark of exceeding toughness, very rich in fibre and well adapted for whipcord, fishing lines and nets, matting, baskets, and paper of fine quality. The plant is found growing plentifully as an underwood in forests and gullies in alpine and subalpine situations. Samples prepared at Melbourne Botanic Gardens.

No. 40. Paper prepared from the "Shining Galingale rush," Cyperus lucidus : *Robert Brown*. Ord. Cyperaceæ.—This plant is widely distributed throughout Victoria. It is found growing on banks of rivers, lagoons, &c., and is gregarious in its habit, thus affording great facility for its collection in payable quantities. The percentage of pulp which this plant yields is very large and of good quality, suitable for making strong packing paper, and with proper machinery would no doubt furnish an excellent printing and writing paper. Samples prepared at Melbourne Botanic Gardens.

No. 41. Paper also fibre made from (large form) Juncus vaginatus : *Robert Brown*. Ord. Juncagineæ.—The tall "Sheathed Rush." Valuable as a paper yielding material. Can be had in large quantities ; gives a large percentage of pulp, suitable for manufacture of strong paper. Found growing extensively on margins of lagoons and water courses in Victoria and other parts of Australia. Sample prepared at Melbourne Botanic Gardens.

No. 42. Paper prepared from Carex pseudo-cyperus : *Linnæus*. Ord. Cyperaceæ. — Found growing on margins of lagoons and water courses often amongst C. appressa. A good paper material but not to be had in large quantities. Sample prepared at Melbourne Botanic Gardens.

No. 43. Paper prepared from the leaves of Gahnia psittacorum : *Linnæus*; var. erythrocarpum. Ord. Cyperaceæ.—A species of sword grass, the leaves of which attain, in favourable situations a length of twelve feet. Found growing chiefly on banks of rivers and creeks, where it can be obtained in very large quantities with ease. Besides yielding material for paper making the leaves can be utilised for common brooms. Indigenous to Victoria and other parts of Australia. Sample prepared at Melbourne Botanic Gardens.

No. 44. Paper prepared from the stems and leaves of a large growing Cyperus, possibly C. vaginatus found growing plentifully on margins of water courses, in subalpine siuations in Victoria. Yields a pulp suitable for manufacture of writing, printing, and packing paper. Sample prepared at Melbourne Botanic Gardens.

No. 45. Paper prepared from the stems and leaves of Lepidosperma elatius : *Labillardière*. Ord. Cyperaceæ.—The tall "Sword Grass." The leaves and stems of this plant grow to a length of nine feet in favourable situations, it is gregarious in its habit and can be had in large quantities. Furnishes a valuable pulp for the making of strong paper and is also adapted for ordinary brooms. Found growing far inland in Victoria principally adjacent to water, in subalpine situations where it attains its greatest perfection. Sample prepared at Melbourne Botanic Gardens.

No. 46. Paper prepared from the stems of Juncus pauciflorus : *Robert Brown*. —A species of small rush found growing on the margins of water courses and lagoons. Furnishes a pulp suitable for fine paper. Sample prepared at Melbourne Botanic Gardens.

No. 47. Paper prepared from Poa australis. Ord. Gramineæ.—This rigid erect growing grass is to be met with in various forms, throughout Victoria and other parts of Australia; chiefly on banks of streams and in marshy places, where it attains a considerable length and could be collected in payable quantities. It furnishes material for a good strong paper. Sample prepared at Melbourne Botanic Gardens.

No. 48. Paper prepared from the bark of Eucalyptus obliqua : *L'Heritier*. Ord. Myrtaceæ.—"The Stringy Bark tree" of the colonists. The fibre producing properties of the bark of this tree are extraordinary it is to be had in almost unlimited quantities in many parts of Victoria and Tasmania and also in the colony of South Australia. The tree grows to an immense height with a diameter of stem of ten feet or more, and clear of branches for a considerable distance. The bark peels readily and is extensively employed by the settlers for roofing their habitations, &c. Although too harsh in itself to make good paper, mixed with other material it can be utilised with advantage in the manufacture of several kinds of paper. The pulp bleaches well and becomes from a rich tan color to yellowish white. The fibre might also be used for some kinds of rope, and for stuffing. The tree also yields a gum possessed of considerable astringent properties. Sample prepared at Melbourne Botanic Gardens.

No. 49. Paper from the bark of Eucalyptus fissilis : *F. von Mueller*. Ord. Myrtaceæ.—"Messmate" of the settlers. A large growing timber tree allied to E. obliqua and closely resembling it in appearance. The remarks upon the uses of the bark of the preceding apply equally to this. Sample prepared at Melbourne Botanic Gardens.

No. 50. Fibre, also paper, prepared from the bark of Hibiscus splendens : *Fraser*. Ord. Malvaceæ.—The "Hollyhock-tree" of Queensland and New South Wales. A beautiful shrub or small tree attaining a height of twenty feet or more, very pubescent, and bearing large rose-colored or deep pink flowers, resembling a hollyhock in size and appearance. Bark very rich in fibre, suitable for fishing lines, cordage, paper, &c. Sample prepared from plant in Melbourne Botanic Gardens, where it grows quickly.

No. 51. Paper also fibre prepared from the bark of Hibiscus heterophyllus : *Ventenat*. Ord. Malvaceæ.—A tall-growing shrub indigenous to the colonies of Queensland and New South Wales but of quick growth in Victoria. Bark rich in fibre suitable for a variety of purposes. Samples from plant grown in Melbourne Botanic Gardens.

No. 52. Fibre prepared from the leaves of Dracæna Draco : *Linnæus.*—The famous " Dragon Tree " of Teneriffe. Sample prepared from plants growing in Melbourne Botanic Gardens, where it is now thoroughly established. Fibre strong and flexible but the tree is of slow growth.

No. 53. Fibre prepared from the leaves of a species of Astelia from New Zealand. The leaves of this plant grow to a length of four feet, and are rich in fibre suitable for ropes, cordage, paper, &c. Sample prepared from plants growing in Melbourne Botanic Gardens, where it is of quick growth and of robust habit.

No. 54. Fibre prepared from the leaves of Yucca filamentosa : *Willdenow.*—The thready " Adam's Needle." A native of Virginia, North America but attaining great perfection in Victoria. Sample prepared at Melbourne Botanic Gardens.

No. 55. Fibre prepared from the leaves of the " Dwarf Palm Lily." Cordyline Pumilio : *Hooker, fil.* Ord. Liliaceæ.—"Ti-rauriki " of the natives. The leaves of this interesting species of Cordyline grow to a length of three feet or more and yield an abundance of fibre of long staple, suitable, for ropes, mats, baskets paper, &c. The plant is of quick growth in Victoria. Sample prepared at Melbourne Botanic Garden.

No. 56. Paper prepared from the leaves of Lepidosperma gladiatum : *Labillardière.* Ord. Cyperaceæ.—The Coast " Sword Rush." This plant which can be obtained in large quantities, on barren sandy places, almost everywhere, along our coast line furnishes one of the best materials for paper, out of the many with which our colony abounds. Attention was called specially to this plant as a paper material several years ago, by Mr. Cosmo Newbery, who exhibited paper made from this plant and other plants of a similar nature at former exhibitions, which attracted much attention at the time. Sample prepared at Melbourne Botanic Gardens.

No. 57. Paper prepared from the leaves of Xerotes longifolia : *Robert Brown.*—The "Tussack Grass " of the colonists. This plant is widely dispersed throughout Victoria, especially on plains and open couutry, and in the neighborhood of water. It grows to a considerable length and furnishes a large percentage of pulp, suitable for packing paper. Sample prepared at Melbourne Botanic Gardens.

No. 58. Paper prepared from the leaves and stems of Arundo conspicua : *Forster.* Ord. Gramineæ.—" The Plume Grass " of New Zealand. This plant although indigenous to New Zealand grows very rapidly in Victoria. The leaves and flower stalks yield a good pulp suitable for the manufacture of several kinds of paper. Sample prepared at Melbourne Botanic Gardens.

No. 59. Paper prepared from the leaves and stems of Gynereum argenteum : *Nees.* Ord. Gramineæ.—The " Pampas Grass " of South America. The remarks on the Arundo conspicua (No. 58) apply also to this magnificent species of grass, in every respect. Sample prepared at Melbourne Botanic Gardens.

No. 60. Paper prepared from bark of Melaleuca ericifolia : *Smith.* Ord. Myrtaceæ.—The " Swamp Tea-tree " of Victoria and New South Wales. The lamellar bark of this tree easily detaches itself and cau be had in considerable quantities. It is suitable for making a soft paper and from its absorbent properties well suited for blotting paper. Sample prepared at Melbourne Botanic Gardens.

No. 61. Paper prepared from the bark of Melaleuca genistifolia : *Smith.* Ord. Myrtaceæ.—One of the largest of the Tea-tree family attaining in favourable situations a height of sixty feet. Indigenous to New South Wales, Queensland and North Australia. Bark possessed of similar properties as that of the common "Tea tree" (No. 60) a remark which will apply with equal propriety to most of the Melaleucas. Sample prepared from tree growing in Melbourne Botanic Gardens.

No. 62. Paper prepared from bark of Pittosporum crassifolium : *Banks and Solander*. Ord. Pittosporeæ.—An erect growing shrub or small tree indigenous to New Zealand but of very rapid growth in Victoria where it is used for ornamental hedges. Most of the Pittosporums of New Zealand are possessed of tough barks, somewhat similar to Pimelia axiflora in texture. Sample prepared at Melbourne Botanic Gardens.

No. 63. Paper prepared from bark of Melaleuca squarrosa : *Smith*. Ord. Myrtaceæ.—"Yellow Wood" of some districts of Victoria. This species grows to a considerable height in parts of Gippsland. It is also found in the colonies of New South Wales, South Australia and Tasmania. Sample prepared at Melbourne Botanic Gardens.

No. 64. Fibre prepared from the bark of Plagianthus betulinus : *A. Cunningham*. Ord. Malvaceæ.—The "Lace Bark" or "Ribbon tree" of New Zealand. A graceful tree attaining a height of seventy or eighty feet. Bark of a beautiful lace-like texture and very strong, suitable for fishing lines and nets, cordage, mats, baskets and paper. It is of very quick growth in Victoria. Sample from plants grown in Melbourne Botanic Garden.

No. 65. Fibre prepared from the bark of Sterculia lurida: *F. von Mueller*. Ord Sterculiaceæ.—A tree of large size, somewhat resembling S. acerifolium. It is a native of New South Wales, but succeeds well, in Victoria. Its bark is valuable as a fibre material, suitable for making mats, baskets, ropes, paper, &c.; it is easily prepared by a steeping process. Samples prepared from plant grown in Melbourne Botanic Gardens.

No. 66. Paper made from Conferva spc.?—This material can be had in enormous quantities, even in the immediate neighborhood of Melbourne. It completely covers the surface of the various lagoons throughout the colony and is very easily gathered and converted into a good strong paper, fit for packing purposes. Samples prepared at Melbourne Botanic Gardens.

No. 67. Fibre prepared from the leaves of Yucca aloifolia.—The Aloe leaved "Adam's-Needle," a native of South America. It succeeds admirably in Victoria and is of moderately quick growth. Sample prepared from plant grown in Melbourne Botanic Gardens.

No. 68. Fibre prepared from Cladium radula: *Robert Brown*.—"The Black Reed " or " Cutting Grass." A tall sedge-like grass, on wet land found growing extensively to a height of three to four feet. It is used by brickmakers as a covering material, and by the settlers as a thatch for their houses, for which purpose it is well adapted. The fibre when properly prepared is strong and of good quality. The plant is a valuable paper material and can be had in any quantity. Indigenous to Victoria and many other parts of the Australian continent.

No. 69. Fibre prepared from Lepidosperma flexuosum : *Robert Brown*. Order Cyperaceæ.—"The slender Sword Rush " known locally as the "Mat Grass." This plant is found plentifully on low-lying swampy ground in the Dandenong district and many other parts of the colony of Victoria. Like most of the genus it furnishes a valuable paper material and also a good strong fibre. It is made into baskets, &c., by the aborigines.

No. 70. Fibre from leaves of Diauella longifolia : *Hooker, Fil.* Order Liliaceæ.— This plant was formerly in great repute amongst the aborigines for basket making, fishing lines, &c. It cannot however be had in payable quantities. Indigenous to Victoria and many other parts of the Australian Continent.

List of Gums, Resins, and Barks.

No. 1. Resin from Eucalyptus fissilis : *F. von Mueller.* "The Messmate."—This substance is obtainable in large quantities, both in a liquid and solid state. Its properties are akin to those of Gum Kino and is sometimes used as a substitute for that important article. Indigenous to Victoria, Tasmania, and New South Wales.

No. 2. Resin from Eucalyptus viminalis : *Labillardière.*—In some districts, known locally as the "Weeping Gum," and "Box-tree," and also as the "Manna" and "Peppermint Gum." The properties of the resinous matter which this tree exudes, in large quantities, are similar to those of E. fissilis, a remark which applies equally to the gum obtained from most of the Eucalypts. The tree is indigenous to Victoria, New South Wales, South Australia, and Tasmania.

No. 3. Resin from Eucalyptus leucoxylon : *F. von Mueller.*—Known locally in the Dandenong ranges Victoria, as the "Milk-white Gum" and in New South Wales as "Iron Bark" and "Black Mountain Ash." It is also found extending over a large tract of country in the colony of South Australia.

No. 4. Resin obtained from Eucalyptus amygdalina : *Labillardière.* — The narrow-leaved "Peppermint-tree" of the lowlands, "Stringy Gum" of the mountains. This tree is extensively distributed throughout Victoria, New South Wales, and Tasmania. It has been known to attain a height of 450 feet and a diameter of 30 feet.

No. 5. Resin from Eucalyptus obliqua : *L'Heritier.* "The Stringy Bark."— This tree grows to a great size, forming immense forests in Victoria, South Australia, and Tasmania.

No. 6. Gum from Panax sambucifolius : *Sieber.* "The Elderberry Ash."—This tree exudes a transparent gum ; especially during the summer months. It can be obtained in considerable quantities. The tree is found inhabiting moist forest gullies, principally in high altitudes in the colonies of Victoria, New South Wales, and Tasmania.

No. 7. Gum from Acacia pycnantha : *Bentham.* "The Golden Wattle."—This substance might be used as a substitute for gum arabic. It is obtainable in considerable quantities. The tree is plentifully distributed throughout Victoria and South Australia.

No. 8. Gum from Acacia dealbata : *Link.* "The Silver Wattle."—Widely distributed throughout the colonies of Victoria, New South Wales, and Tasmania. It bears a close affinity to A. decurrens, the common Wattle, and is found invariably on banks of watercourses. Gum similar to that of A. arabica, and may be obtained in large quantities.

No. 9. Gum from Acacia decurrens : *Willdenow.* "The Common or Black Wattle."—The gum exuded by this tree is similar in every respect to that of A. dealbata. The tree is distributed throughout the colonies of Victoria, New South Wales, South Australia, and Tasmania, in some places springing up with great rapidity, and forming dense impenetrable scrubs.

No. 10. Gum similar to Sandarac from Callitris rhomboidea : *Robert Brown*. "The Native Cypress."—This tree is found in the colonies of Victoria, New South Wales, Queensland, and South Australia.

No. 11. Gum Sandarac from Callitris Gnnnii : *Hooker*. "The Native Cypress" of Tasmania. — Sample obtained from plants growing in Melbourne Botanic Gardens.

No. 12. Resin from Synearpia lanrifolia. The New South Wales "Turpentine-tree."—Sample obtained from plants growing in Melbourne Botanic Gardens. This tree is of tall and slender habit and is indigenous to the colonies of New South Wales and Queensland.

No. 13. Gum from Grevillea robusta : *A. Cunningham*. "The Silky Oak."— The exudations from this tree, appear to be distinct in character from any other of the native woods. It is of a pale yellow color and very teuacious. The tree is in-digeuous to New South Wales and Queensland. Sample obtained from trees grow-ing in Melbourne Botanic Gardens.

No. 14. Gum from Sterculia diversifolia : *G. Don*. "The Victorian Bottle Tree."—Found also in New South Wales and Queensland. The substance which this tree exudes, in large quantities, would no doubt form a good Tragacanth.

No. 15. Gum Sandarac from Callitris robusta : *Robert Brown*. "The Murray Pine."—Indigenous to Vietoria, bordering the river Murray, and throughout all other parts of the Australian continent. Often growing on barren sandy wastes.

No. 16. Gum obtained from Hakea gibbosa : *Cavanilles*.—A tall growing shrub or small tree indigenous to New South Wales. Sample from plant growing in Melbourne Botanic Gardens.

No. 17. Gum from Corynoearpus lævigata : *Linnæus*. "The New Zealand Laurel."—A beautiful glabrous, leafy tree, of pyramidal habit, growing in favourable situations to a height of forty feet or more. Native of New Zealand. Attains to great perfection in Victoria where it is of moderately quick growth. Sample from plants growing in Melbourne Botanic Gardens.

No. 18. Resin from Araucaria Cunninghamii : *Aiton*. The "Moreton Bay Pine."—A noble tree, native of Queensland and New South Wales. Samples obtained from trees growing in Melbourne Botanic Gardens.

No. 1. Bark of Atherosperma moschata: *Labillardière*. "The Victorian Sassafras."—This bark is highly esteemed for its aromatie and astringent proper-ties, and may be had in considerable quantities. The tree is indigenous to Victoria and Tasmania.

No. 2. Bark of Melaleuca ericifolia : *Smith*. "The common Swamp Tea Tree." —Useful as a paper material. It might also be utilised in the manufacture of hats suitable for hot climates. It is obtainable in vast quantities in many parts of Victoria. Sample of paper, also exhibited, made from this bark. New South Wales and Tasmania.

No. 3. Bark of Melaleuca squarrosa : *Smith*. "The Victorian Yellow Wood." —In Gippsland, attaining a height of sixty feet, its lamellar bark, which can be stripped off in large sheets, serves for thatching and can be utilised in a similar manner to that of the preceding. The tree or shrub is extensively distributed over Victoria, New South Wales, South Australia, and Tasmania. Sample of paper prepared from this bark also shown.

No. 4. Bark of Melaleuca genistifolia : *Smith.*—A very large growing "Tea-tree." A native of New South Wales, Queensland, and North Australia. Sample from tree growing in Melbourne Botanic Gardens, paper prepared from this bark also exhibited.

No. 5. Bark of Eucalyptus obliqua : *L'Heritier.* The "Stringy Bark."—Used extensively by the settlers as a thatching material. Sample of paper prepared from this bark also shown.

No. 6. Bark of Eucalyptus fissilis : *F. von Mueller.* "The Messmate."—The bark of this tree is also used extensively as a thatching material. Sample of paper also exhibited.

No. 7. Bark of Acacia pycnantha : *Bentham.* "The Golden Wattle."—This bark is used extensively in tanning leather, its astringent properties being very great. It might also be converted into a strong packing paper. The tree is distributed over the colonies of Victoria and South Australia.

No. 8. Bark of Acacia decurrens : *Willdenow.* "The common Black Wattle."— Bark extensively used for tanning leather; might also be utilised for paper making. Extensively distributed through the colonies of Victoria, New South Wales, Queensland, South Australia, and Tasmania.

No. 9. Bark of Acacia dealbata : *Link.* "The Silver Wattle."—Considered by some authorities to be synonymous with A. decurrens, var. mollissima. But according to Hooker, it is sufficiently distinct to be separated from the above. This bark is similar in its astringent properties to A. decurrens, and is extensively used for tanning ; it is also available for paper.

No. 10. Bark of Pimelia axiflora : *F. von Mueller.*—This shrub is extensively distributed through the fern gullies, and humid forest valleys of Victoria and New South Wales. Its bark supplies a valuable fibre for whip cords, paper, &c. Samples of paper and fibre from this plant also shown.

No. 11. Bark of Quercus suber : *Linnæus.* "The Cork Oak" of South Europe and Northern Africa.—This tree is now thoroughly established in the Melbourne Botanic Gardens and many other parts of Victoria. Sample obtained from tree growing in Melbourne Botanic Gardens.

Dr. Sturt of Emerald Hill has kindly supplemented our list of exhibits, by presenting two bottles of a preserve made from the "Kai Apple" (Aberia Caffra), a native of South Africa, but thriving well in Victoria. The plant bears enormous quantities of fruit of a bright yellow color in size and shape somewhat like the Golden Pippin apple. This shrub is very suitable for hedges.

List of Dyeing Materials, &c., for Exhibition purposes.

No. 1. Dye, obtained from bark of Pipturus propinquus. The "Queensland Grass-cloth-plant."—This bark gives—under different treatment—several beautiful shades of brown. Samples of paper and fibre prepared from this bark also exhibited

No. 1A. Piece of woollen cloth, dyed with extract from bark of Pipturus propinquus. Mordanted with *Sulphate of Iron.*

No. 1B. Piece of woollen cloth also piece of silk dyed with bark of Pipturus propinquus.

No. 1c. Piece of calico stained with liquor obtained from bark of Pipturus propinquus.

No. 1d. Piece of woollen cloth, also piece of silk dyed with bark of Pipturus propinquus. Mordanted with *Chloride of Tin*.

No. 2. Piece of woollen-cloth, also piece of silk dyed with bark of Dais cotinifolia, South Africa, Mordanted with *Sulphate of Iron*. Samples of paper prepared from this bark also exhibited.

No. 2A. Piece of woollen-cloth, also piece of silk dyed with bark of Dais cotinifolia.

No. 2B. Piece of woollen cloth, also piece of silk dyed with 'bark of Dais cotinifolia. Mordanted with *Chloride of Tin*.

No. 2c. Piece of woollen cloth, also piece of silk dyed with bark of Dais cotinifolia. Mordauted with acetate of Iron.

No. 3. Piece of woollen-cloth dyed with the tubers of the "Sundew," Drosera spe.—The roots of this beautiful little plant possess similar properties (as a dyeing material) to the bark of the "Queensland Grass-cloth Plant," Pipturus propinquus (No. 1).

No. 4. Dye, obtained from the bark of Pimelia axiflora (Currijong of the Aborigines). Qualities not known at present. Samples of paper and fibre prepared from this bark also exhibited.

No. 4A. Piece of woollen cloth also piece of silk dyed with bark of Pimelia axiflora ("Currijoug" of the Aborigines).

No. 5. Dye, obtained from bark of Laportea gigas. The "Tree Nettle" of New South Wales and Queensland.—This bark apparently possesses properties exactly similar to those of the bark of Pipturus propinquus (No. 1). Sample of fibre prepared from this bark also exhibited.

No. 5A. Piece of woollen cloth also piece of silk dyed with bark of Laportea gigas. The "Tree Nettle" of New South Wales and Queensland.

No. 6. Dye, obtained from the husks of Sterculia diversifolia. "The Victorian Bottle-tree."—Quality at present unknown. Samples of fibre and paper, also *fat-acid* obtained from the seeds, exhibited.

No. 6A. Piece of woollen cloth also piece of silk dyed with husks of Sterculia diversifolia. The Victorian "Bottle-tree."

No. 7. Oleo-resin, obtained by a boiling process, from seeds of Pittosporum undulatum, "The Native Laurel," Victoria.—The properties which these seeds possess are not sufficiently known at present to admit of a definite opinion being expressed as to their economic value. There is no doubt however that with proper appliances a valuable resin might be obtained from them in payable quantities, on account of the abundance of fruit which the tree produces. Sample of polished wood, from branch of this tree also exhibited.

No. 8. Fat-acid, obtained from the seeds of Sterculia diversifolia, "The Victorian Bottle-tree."—These seeds would, no doubt yield by expression or otherwise, an oil or fatty-matter of considerable value. Samples of fibre and paper prepared from the bark of this tree also exhibited, also a dye obtained from the husks of the seed.

By Authority: GEORGE SKINNER, Acting Government Printer.

www.ingramcontent.com/pod-product-compliance
Lightning Source LLC
Chambersburg PA
CBHW022014190326
41519CB00010B/1523